T3-BED-162

Managing Technical Services
in the 90's

Managing Technical Services in the 90's

76622

Drew Racine
Editor

The Haworth Press, Inc.
New York • London • Sydney

Managing Technical Services in the 90's has also been published as *Journal of Library Administration*, Volume 15, Numbers 1/2 1991.

The Haworth Press, Inc., 10 Alice Street, Binghamton, NY 13904-1580
EUROSPAN/Haworth, 3 Henrietta Street, London WC2E 8LU England
ASTAM/Haworth, 162-168 Parramatta Road, Stanmore, Sydney, N.S.W. 2048 Australia

Library of Congress Cataloging-in-Publication Data

Managing technical services in the 90's / Drew Racine, editor.
    p. cm.
    "Has also been published as Journal of library administrtion, volume 15, numbers 1/2, 1991"—T.p. verso.
    Includes bibliographical references.
    ISBN 1-56024-166-7 (alk. paper)
    1. Processing (Libraries)—Management. 2. Library administration. I. Racine, Drew.
Z688.5.M29   1991
025'.02—dc20

                                                   91-24386
                                                    CIP

Managing Technical Services in the 90's

CONTENTS

The Role of Librarians in Bibliographic Access Services in the 1990's

Jennifer A. Younger

ABOUT THE EDITOR

Drew Racine is Deputy Assistant Director at the University of Texas at Austin. He was previously Program Director of Research Libraries at OCLC, Inc., and Assistant Director for Technical Services at the University of Missouri. Mr. Racine is a member of the American Library Association, ACRL, ALCTS, and LAMA.

Introduction

Change. As a verb it describes a physical and biological imperative. As a noun it is sometimes a scapegoat on which we can lay
blame for a multitude of problems. It seems that one cannot read an
article in the professional literature these days without being told
that change or rapid change or continuous change or an escalating
rate of change or another variation on this theme requires modifications in attitudes and behaviors on the part of librarians. That this
truism is so often stated is not a problem (if one doesn't consider
repetition to be a problem), but what may be a problem is that
change is still too often described as if it were an aberration.

Is this issue of the *Journal of Library Administration* about
changes in libraries and technical services? Yes. Is it about managing technical services in the '90s? Yes again. The former is significant; the latter is less so. For, while we may be tired of reading and
hearing about change, we know from coping with changes in every
facet of our lives that change is in fact a constant, not an aberration,
that it has always been so. In 1845 in *An Essay on the Development
of Christian Doctrine*, John Henry Newman wrote: "In a higher
world it is otherwise, but here below to live is to change; to be
perfect is to have changed often." We need to internalize the fact
that the normal course of events is changing; that change is not
threatening some sacred status quo, but is creating whatever status
quo exists; that the concept of time without change is a false construct. This is neither aberration nor crisis; this is life.

The function of technical services remains what it has always
been in the broad sense, that is, to produce products and services to
satisfy the information needs of "the public." But many fundamental changes have taken place in the library. Thanks to technological
advances, we now have more and better methods to deliver more
varied products and services. Though it sounds quite paradoxical, the
concept of a "library without walls" is becoming concrete reality.

1

Both library staff and "the public" are more diverse in many ways, are more aware of the possibilities, and are more assertive about their needs. The more traditional "POSDCORB" functions of the manager have taken second place (or lower) to more highly regarded managerial and organizational qualities as entrepreneurship, innovation, flexibility, and taking risks. Governmentally funded entities, such as libraries, receive a smaller and smaller portion of their funding from the governments, and as a result, fund-raising (or development) has become an increasingly accepted method of obtaining basic operating funds — competition for dollars is more keen.

New technologies, new models of management, new organizational paradigms, new social and psychological approaches necessitate differing actions and responses for all persons who manage, whether the area managed be highly technological, highly interpersonal, or both. A plethora of respected librarians (see for example, Gapen (1989); Graham (1990); Hoadley and Corbin (1990); Woodsworth et al. (1989)) have expressed their ideas about how changes in society, in the information industry, and in libraries require iconoclastic thinking and acting on the part of librarians to take maximum advantage of the opportunities that are present to better the profession, the professionals, and services to information seekers. These articles hope to further this kind of thinking.

This issue of the *Journal of Library Administration* focuses on managing functions typically associated with technical services. But actually, technical services is not a place or an outlook on life and work, but rather is shorthand for the functions commonly performed in traditional "technical services" divisions. Most of the authors have worked primarily in academic libraries. The articles are more speculative than research, and are therefore written both for managers of those functions and for those who are concerned with how those functions are managed. Both current and aspiring librarians were kept in mind as the articles were being written. Much of what is written will be valuable beyond those interested in technical services because, as is pointed out either explicitly or implicitly in several of these articles, the mission and the management of technical services are no longer very different from the missions and management of other library units, and, in fact, not very different from the mission and the management of the library as a whole.

William Gosling's article sets the stage for the other articles. He describes the current state of affairs in technical services and foreshadows many of the management issues that the other authors discuss in more detail. His contribution identifies issues, potential problems, and opportunities for further research and thinking.

Delmus Williams "ruminates" about recent changes in technical services, their effects on the library, and needed changes in the near future. Williams believes that some of the cloudiness of the recent past due to unsurety has cleared and that it is time to forge ahead, integrating technology and changes made possible by technology into the library of the future. In his speculations, Williams makes it clear that more changes must come about and that technical services is pivotal in this process. He focuses on staffing in the library and in technical services, the mission of the library, organizational structures, managerial styles and comfort, and the still anticipated "technology dividend."

Arnold Hirshon describes two ways of looking at technical services in relation to the rest of the information universe. The future value of technical services, and indeed of the library, depends on changing the world view from which technical services and the library view themselves and their roles. Library administrators must look outside their organizations to form a broad vision for the future of their libraries and then must focus on specifics to realize that vision. Hirshon presents a planning matrix to help administrators and managers identify problems and develop solutions.

Sharon Walbridge discusses new partnerships between librarians who work in technical services and those who work in public services. The partnerships have been enabled by technological changes. Both the changes and the resulting staffing flexibilities can mean significant improvements for the users of the libraries and for those who work in them.

Wayne Perryman recounts the major changes in the recent past affecting the ways information is created, stored, accessed, and delivered. As a result of these changes, libraries have opportunities to provide access to information in a variety of ways, some old, some new. Selecting from among these methods has ramifications for all facets of library operations down to the most basic level of service. Perryman discusses the evolving information infrastructure and libraries' evolving attitudes about acquiring and communicating information, and calls for a common vision to be held by librarians.

Olivia Madison takes a more micro-level view of the ways that new technological capabilities have expanded the abilities of libraries to provide bibliographic and physical access to information. As the distinction between these two facets of information seeking becomes less and less — in both the physical and temporal dimensions — due to the technology, libraries must adapt and take advantage of these capabilities or risk wrack and ruin.

Kenneth Bierman applies many of the lessons of the other authors to public libraries. While libraries of all types have basic technical services functions in common, there are significant differences between academic and public libraries in how these functions are carried out. But once these differences are pointed out, often the solutions again converge.

Finally, Jennifer Younger examines management of the bibliographic control function in libraries. Bibliographic control is one of the major functions of libraries in terms of its importance and the resources expended on it. Due to the online catalog, bibliographic control is no longer subject to many constraints formerly applied to it, internally or externally. While discussions of the difficulty of attracting and retaining excellent librarians in bibliographic control work are heard from many directions, Younger describes new roles and responsibilities for librarians in management, staff development, technical analysis and design, setting and interpreting standards, research, publishing, teaching, intellectual leadership and modeling, and other functions that are necessary for excellent programs of bibliographic control as well as for participation in the full range of services offered by the library. Younger makes a very good case that, rather than being a wasteland, careers in bibliographic control can be among the most challenging and rewarding in the library.

There are common themes across these independently written articles. The need for flexibility in staffing and organizational structures is one thing you will readily identify. Looking at the traditional in new ways is another. Convergence and union is another. In spite of the myriad and very major obstacles that librarians must surmount and which sometimes seem to be grinding them down, there is quite a bit of optimism in these articles. There is faith that librarians can manage change and improve information services in real terms. If one can generalize from these eight articles, one might state that librarians are progressing into the stage of acceptance of new technologies in which new things are done and changes at the very foundations of libraries are being made. As we move along the road to Newmanian perfection, I am glad that I am a librarian at this time.

Drew Racine

REFERENCES

Gapen, D. Kaye. "Transition and Change: Technical Services at the Center." *Library Resources and Technical Services* 33 (July 1989): 285-296.

Graham, Peter S. "Electronic Information and Research Library Technical Services." *College and Research Libraries* 51 (May 1990): 241-250.

Hoadley, Irene B. and John Corbin. "Up the Beanstalk: An Evolutionary Organizational Structure for Libraries." *American Libraries* 21 (July/August 1990): 676-678.

Woodsworth, Anne, et al. "The Model Research Library: Planning for the Future." *Journal of Academic Librarianship* 15 (July 1989): 132-138.

Managing New Technological Products and Services in Technical Services: The Current Scene

William A. Gosling

Today's technical services manager is facing a variety of demanding challenges. The working environment is radically different from what is was even five years ago. Most libraries have moved from a card-based access system to online public access catalogs and related service features. These catalogs are based on computer support systems that librarians may or may not oversee. The manager in this setting must be prepared to deal with an ever-broadening array of issues and management situations while striving to direct these evolving operational services in support of the knowledge-based society.

Whether managing at the upper or middle supervisory level, many of the same skills are required of the technical services manager. Senior managers most likely have learned automation applications through on-the-job training, continuing education, and participation in conference seminars and programs. Younger librarians and higher ranked paraprofessionals are bringing to the job a greater knowledge of automated applications learned as part of their basic education, enabling them to manage the automated operations more knowledgeably in library technical services. It is essential that today's managers have a solid foundation not only in the basic machine-readable cataloging (MARC) structure, which is the foundation of most library bibliographic access record systems, but in the

William A. Gosling is Assistant Director for Technical Services at the University of Michigan, Ann Arbor, MI.

automation elements — hardware, software, and networking. Without such preparation, the manager is seriously handicapped. While general management expertise will enable basic administration of many technical services operations, it is becoming more and more critical that all managers have a broad knowledge of automated systems to be successful in their work.

For many libraries, the card catalog has been replaced by an online public access catalog of machine-readable records based on the standardized MARC format. Cataloging records are derived from a variety of bibliographic sources: some may come directly from Library of Congress products on tape, CD-ROM, or print; others may be secured from specialized vendors, who provide machine-readable records, often with preprocessed authority records; or records may be derived from major bibliographic utilities which specialize in support of a variety of library bibliographic services, including provision of cataloging records for local online systems, automated interlibrary loans, and regional shared databases. As recently as five years ago, many libraries continued to produce catalog cards which were filed into extensive manual catalogs, using these same information sources. These card catalogs have been "frozen," if not physically replaced by online records in most libraries.

Patrons and staff can manipulate this machine-readable bibliographic data in a variety of ways, both within the library and from remote access terminals. This remote capability has greatly expanded access to the library's collections and, through local and regional networking, has opened up an institution's resources to a much wider community. Expectations of users, administrators, and staff have been raised, in turn, as to the accuracy of these records and the delivery mechanisms within the library to supply the desired materials or data.

Many libraries have been able to secure funds to complete retrospective conversion of the bibliographic records that previously made up their card catalogs. Others have only current cataloging data online, with large retrospective files yet to be converted. As libraries complete their retrospective conversion and add their records to the national bibliographic utilities, other libraries benefit from a higher rate of matches when they initiate their retrospective conversion projects.

Initially, libraries replicated their card environment in the online public access catalog. Now, automated systems provide patrons with new methods of accessing bibliographic data. Where the card catalog limited access points to author, title, series, and subject, online keyword boolean searches now make major portions of the entire bibliographic record searchable by the patron. This brings with it the expectation that information throughout the bibliographic records will be accurate to facilitate retrieval, as contrasted with the card file where only a single access point required correct spelling for the patron to be successful in finding desired information.

Many libraries have expanded access to information beyond contents of the card catalog in the electronic version. Coverage includes not only a record of the holdings of a given institution, but citations to multiple external information datasources. Libraries have added records from other institutions, such as the Center for Research Libraries, to the local catalog. Groups of libraries have formed joint regional catalogs that support a local area network, often of different types of libraries, serving multiple community groups. It is not uncommon to find a local public library networked together with area academic and special libraries, each benefiting from the resource sharing that such cooperative access provides. Sharing a common database, these libraries are able to realize some economy of operation through the use of one centralized computer facility, one set of software, and the maintenance of single bibliographic records with multiple holdings statements. An alternative configuration may consist of linking together several libraries' local catalogs and having them function as a single online server. In either cooperative arrangement, uniquely held items can be readily located by the patron from one terminal rather than having to travel from library to library.

These new catalog configurations also may place new demands upon the technical services staff. Not only must the unit head be concerned with local practices, the manager must be able to cooperate in coordinating technical operations and file creation among multi-institutional needs.

These new means of accessing library collections, in turn, provide opportunities for rethinking the creation of and securing of bibliographic records. The national utilities provide the majority of many institutions' records while others rely on commercial vendors for bibliographic citations. Some libraries draw on regional cooperative union catalogs as noted above. Specialized files may be secured en bloc from vendors, such as for large microform sets, government documents, and machine-readable datafiles. Original cataloging continues to be produced locally for less commonly held titles, especially foreign language and regional publications. Addition of these records to the national databases provides access to the more unique materials within the national library community.

With the introduction of more computerized applications in the production and dissemination of "published" materials, librarians must be able to handle an increased quantity of data and information packaged in a rapidly growing range of new formats. These new formats are additions to rather than replacements for most of the traditional print material. Technical services units continue to receive a large volume of books and journals published on paper for processing. Proliferation of formats is adding to the workload with the repackaging of some of this information, such as journal indexes, on CD-ROM or in online machine-readable datafiles. New materials also are appearing in their first and only edition in electronic form. This requires that cataloging records be created for software, compact disks, a variety of video and film materials, and machine-readable datafiles for inclusion in the online public access catalog. In turn, these changes increase demands for catalogers with new skills and supervisors who are able to plan for and provide equipment support to enable proper processing of such materials.

In many settings, the access catalog is becoming a file which describes the holdings of the library as well as a wide array of information sources not owned by it but available through the library to the patron. These may include full text files, statistical datafiles, software in a microcomputer center, and a range of remote sources now available through local network connections. This introduces a whole new set of challenges for the database manager: how to maintain the database when remote information sources change but staff are uninformed of the alterations; how to describe protocols necessary to access the remote data; how to prepare records in the absence of cataloging standards or models for emerging machine-readable data sources; and who has the information about such sources to enable preparation of an accurate, retrievable record.

Not only has technical services staff shifted rapidly the format in which access to information is made available to patrons within the local library setting, but the equipment used for these processes is significantly different from that of a few years ago. Where once typewriters were heavily used to produce catalog cards, advanced microcomputer workstations are in place to facilitate the electronic transfer of records from one datafile to another. The intermediate step of editing a paper version of records has been replaced by an online review, and the direct passing of bibliographic information from the source file to the local online system is becoming common. Management of authority information is being handled in a variety of ways, with a machine-readable version of the authority records also being passed from the utility to the local system where some variety of automated authority processing is usually performed. Where staff do not yet have in place the mechanisms to make this transfer, records may be written to tape and their processing contracted out to vendors who provide authority processing services. Name, series, subject, and uniform title headings are adjusted to conform to the most current version of Library of Congress and other authority headings created by the library community. Copies of supporting authority records also are written to tape for loading into the local online catalog. These records, in concert with the bibliographic data, form the basis of the online public access catalog.

It is no longer adequate for many institutions to use a single source for current cataloging records. Multiple utilities are consulted, and creative solutions are being sought to acquire records for processing of materials within a given library. Special source files for dissertations, government documents, and large microform set records are frequently utilized to produce local records. While providing analytics for large microform sets continues to be a major initiative, there is beginning to emerge a similar need for analytics for large full-text, machine-readable datafiles such as the ARTFL datafile of French literary and historical works. There is a rapidly growing need to catalog datafiles accumulating in various information centers around the world.

Acquisition operations have benefited as library procedures have become more fully automated and the vendors supplying the material have implemented electronic systems. While many institutions have had automated order tracking and accounting systems in their acquisitions divisions, these are in a state of transition to add online ordering of desired titles, using telecommunication links between the library and the vendor. In addition, greater automation of institutional accounting systems has altered appreciably the methods used to pay for the materials acquired by the library. Accounting activities are easily maintained locally as information is transferred within the institution electronically between the receiving point and the paying point. Vendor information files noting availability of titles from inventory are often supplied to libraries on microfiche or disk to facilitate ordering against known vendor stock. In other cases, online inquiries to vendor inventory datafiles are complementing this acquisition effort.

Serial vendor datafiles are available to provide information about changes in pricing patterns and trends, especially in this era of rapidly escalating journal prices. Such information is becoming essential to larger library systems where annual increases in serial subscriptions may exceed 25% to 35%. By having timely advance information on such journal price trends, collection development and technical services managers are able to predict any major budgetary impacts that these price fluctuations may create. Analysis of local acquisition systems data likewise can provide a detailed profile of the library's ordering patterns. Comparing that profile with the subscription agent's information, library managers can prepare budget projections or initiate serial cancellation projects to balance a budget before, rather than after, the new pricing pattern has exhausted available funds.

Receipt of material has been enhanced through the use of automated systems. Online serial check-in and acquisition receiving systems provide immediate information on receipt of materials and a tracking mechanism as they move through the processing continuum.

Labelling operations, which once required manual keying of the call number on spine labels, have given way to a machine-generated spine label produced from the bibliographic record. This time saving step also ensures greater accuracy of call numbers affixed to the piece in relation to the bibliographic notation.

Catalog maintenance has been enhanced greatly through automated developments. In order to correct a record in the card environment, a staff member would edit or revise the existing record in the utility and generate new cards for filing into the local card catalogs, replacing the old card set. Today, changes to the online file are accomplished efficiently and quickly by machine batch process or by manual interactive editing of the record within the local online catalog. Where previously only a modest number of changes could be made in the course of a year due to the labor intensive nature of the older card-based process, now thousands of changes can be made quickly and economically, often through the use of global change commands. Online maintenance is making it possible to respond to the request for corrections in all elements of the online record in a timely manner and at a rate that would have been impossible in the cardfiles. The result is a more accurate and easily maintained catalog.

One of the major shortcomings that the current technical services unit head must accommodate is the absence of standards of quality, especially in their supporting equipment, and even in some areas of the cataloging rules. As technological advancements emerge, it often takes years before industry standards are established. The field of sound recordings has moved from disk to cassette to compact disk with many variations, especially for the institution acquiring material on an international scale. This is equally true of video and other nonbook products. Compact and optical disk materials are in their infancy as product media, undergoing many refinements and often requiring special players. The collection development staff must be knowledgeable enough to select wisely materials in a field where as many as fifty versions of a given item may be available from which to choose. Often processing staff must create a local original cataloging record to provide access to the unique features of that version. The technical services manager also may be the person who identifies, orders, and oversees installation of the equipment that will support its use.

Certainly, the most demanding element of overseeing technical services operations is the current rapid pace of change. This is occurring both in the materials to be processed in an ever-widening array of formats and the library systems and networks that support their processing and use. In turn, the bibliographic infrastructure must provide access to this broadening range of library materials in the dynamic environment of the online catalog. The time has passed when one would identify a new direction, put in place the planning and budget mechanisms to move forward that initiative, and once there, sustain a new plateau of operations and services. The technical services workplace has become one of rolling change where a new service or product is barely in place before there is a need to accommodate still newer products or upgrades with their parallel demands on fiscal and personnel resources. It may become beneficial to skip certain generations of technological products rather than take each generation in sequence. This is but one of many issues with which the unit head must deal in overseeing the operation.

In the area of personnel, the management of this change is extremely critical as the individual staff member seeks to establish a comfortable work setting. Such an environment is becoming much more difficult to create as constant change produces an instability in the workplace with which the average employee is not comfortable. Staff want to know what they should do, what expectations there are in meeting new production norms, and what to strive for as the new "routine" work. Rapid and frequent changes in automated operations introduce an ongoing need for detailed, frequent retraining programs and the skilled staff to deliver the training successfully. The manager's role includes knowing what changes require training programs, which employees are having difficulty making the latest transition to a new product or procedure, and which operations need to be redesigned to address stress and fatigue caused by long hours at workstations.

We appear to have arrived at a point where many automated systems undergo some adjustment of either software or hardware or both on as frequently as six month cycles. How often can an institution afford to replace its public access software, the hardware supporting the online catalog, or acquire the same data in yet another electronic format with appropriate supporting equipment? The technical services manager is often regarded as the knowledgeable local expert to lead the analysis and initiate the recommendations of when and how to incorporate these generational transitions.

In many institutions, the online catalog is one example of a principal tool for which the software continually is being reviewed by its producers. Revised software means added revenue for the software manufacturer or service firm, stimulating their business and keeping their products current. It is not uncommon to receive updates or revised versions of the software for upgrading the system at six month intervals. For the manager, this brings not only the need to incorporate each of these changes (they usually cannot be passed over) but the need to identify the funds to acquire, install, and maintain them. It is now more than ever a basic necessity to coordinate appropriate training for staff to enable them to master the new features of these enhanced versions as quickly as possible. Therefore, the manager must be knowledgeable not only about changes in the products being received but in the basic hardware that supports automated systems, including microcomputers, telecommunication services, and utility support systems. All of these elements undergo regular change, and it becomes essential to adjust technical services operations to them.

Where once it was sufficient to have access to a telephone line, it is becoming increasingly important to connect to a variety of telecommunication networks serving the information needs of library patrons. Without them, an institution is hampered in its efficient operation and provision of services, not only in technical services but for public service information access as well.

MANAGING

The skills of the manager to function effectively in a modern technical services division require a broad set of characteristics. In his article, "What effective general managers really do," John P. Kotter notes: "They spend most of their time with others. The average general manager spends only 25% of his working time alone, and this is spent largely at home, on airplanes, or while commuting."[1] This description also applies to the technical services manager who must meet often with subordinates, peers, and administrators. In addition, frequent meetings may be necessary with computer center and community counterparts. Knowing which questions to ask to maximize meeting time and to get the most usable information as efficiently as possible is an essential skill for this administrator.

If one does not have information about a format, a product, or a service, one needs to be able to secure details about it rapidly. A broadbased background in technological and computerized applications is fundamental if the manager is to interpret the suppliers' data and secure the best products and services for the institution. There are few situations where the manager has the luxury of an extended period of time to make a decision about advancing to an innovative system or acquiring a piece of equipment. This ability needs to be coupled with sharper business skills as one deals with an evermore complex range of vendors and suppliers, not only of bibliographic information but of the hardware and supporting software to provide the delivery mechanisms for a given library or library system.

While historically it was necessary to keep a "weather eye" on when additional card cabinets would be needed in a given library, the manager now must focus that same "weather eye" on disk support space for the online system. It is essential to have knowledge of the disk equipment and the range of issues associated with securing adequate disk storage for the institution's datafiles. This becomes especially critical in the library that supplies locally machine-readable indexes or full-text products. These products can be mounted with the local catalog, generally in a mainframe or minicomputer environment, using the same search access software as is used for the online catalog. These indexes, however, require significant storage and indexing space. The successful manager must have the ability to plan for sufficient disk space, computing capacity, and terminal support to provide adequate online service.

PERSONNEL

The head of the technical services unit needs to be a leader to help the staff as they adjust to frequent change. As a leader, this individual must have a full command of technical services procedures and of automated systems applications to instill confidence in the employees. Cataloging staff are incorporating expertise in describing machine-readable based products and creating records for datafiles. These entries require descriptions of their special features such as how to access a particular datafile, what kinds of peripherals are needed, and whether special accounts are required to access the given information source. Their leader must be able to direct them successfully in using the most productive means to create these new bibliographic products.

Another concern in managing this ever-burgeoning mass of information is to balance the need for production with a humanistic working environment for staff, realizing the most efficient operation in the process. Upper level administrators are applying more stringent business applications to institutional operations. With these growing expectations, the manager is caught between ensuring compliance with budget and organizational constraints and the need to provide a productive work environment for employees. If not managed properly, workplace tensions may build rapidly, and the effectiveness of the manager could be greatly undermined with the possibility of mounting stress leading to a decline in staff production.

Interpersonal skills are becoming of greater importance in libraries. Institutions employ a more diverse workforce with a greater variety of staff expectations. Training often is useful in fulfilling the leader's role in this dynamic setting. These conditions are placing greater personal demands upon the manager of technical services operations at the very time the need to coordinate implementation of rapidly changing technological applications is growing.

This situation is further compounded by higher patron expectations, especially in organizations where individuals or academic departments can adopt more speedily innovative automated applications for information handling than can be implemented by the library staff. The manager needs to meet with the patron and to respond by describing succinctly what the library realistically can provide.

The successful manager must be able to interact with colleagues in administrative and public service areas of the library, as well as with a variety of supporting vendors. The ability to lead discussion across units to a common resolution of an issue is critical. Library administration is increasingly a joint effort of many unit supervisors. Technical services leaders often gain technical expertise that is extremely useful to the entire organization in providing automated services. However, this introduces another challenge for the management of technical services in the automated environment as it potentially requires additional training in new service skills. It requires protecting worktime to get the primary task done in a situation where new demands are being placed upon a limited staff resource.

BUDGETING

Budgetary skills are another essential ingredient for success in library management. Demands for resources can shift radically, and as an institution experiences a shift from processing book materials to processing electronic datafiles, different levels and skills of personnel may be required. Partly through reallocation, partly through budget expansion, and more often through staff retraining, the technical services manager must employ creative mechanisms to keep materials moving through the processing workflow. Large book backlogs of the past are not acceptable in the electronic age even though in-process records can provide access to materials for patrons from the time of receipt. More libraries are processing materials as expeditiously as possible, in some cases relying on briefer bibliographic descriptions where cataloging copy does not exist from other sources. An emerging philosophy, as reflected in Sheila Intner's recent article, "Copy Cataloging and the Perfect Record Mentality,"[2] notes with the ever-expanding amount of material to be acquired, described, and physically prepared for patron use, a more expeditious means of creating information access must be developed. Purchase of existing records, such as UMI's dissertation abstracts for local dissertation records, becomes more economical than creating or converting records locally. Batch purchase of analytics for large microform sets and machine-readable datafiles is being adopted widely. Some institutions continue to view bibliographic record processing as a local application, editing any and all records to rigid local standards. More and more institutions, however, are recognizing that a verified national record transferred into the local system does not warrant this extensive and expensive revision. Staff dollars are declining in many libraries while materials budgets continue to expand to meet price increases and publication expansion. As a result of reallocation and streamlined operations, staff time needs to be reassigned for processing of the newer formats.

THE INTEGRATED DATABASE

The online public access catalog is now the major information source for the average library. It is an integrated system supporting all functions from ordering and receiving to cataloging and labelling, as well as circulation and patron access. As noted above, it reflects not only the library's collection contents but a variety of other information sources that are available to patrons with instructions on how to access them. One needs to construct this tool so that unlike the card catalog before it, it does not require a major conversion effort later. Many consider the online catalog to be the core of library functions, the most critical product from contemporary technical services operations. It is this data which now supports all library functions. This data which also will move forward to the next generation of hardware and software supporting the OPAC. The integrity and consistency of the data within that online catalog become essential to minimize the cost associated with its transferability to new environments as they become available. Consistency within the data structure is central for the maintenance of the database through automated control and standardization of record format and cataloging practices. Movement of that data to a succeeding system will require major data-mapping efforts, tasks which will be extremely costly if the integrity of the present file is not monitored carefully.

Some institutions already have realized this pitfall as they worked to bring together information from multiple machine-readable files to create their current OPAC. They have sought a machine solution to creating portions of the online catalog; more machine solutions will continue to be sought as a fuller range of information in electronic format can be manipulated in a variety of ways to produce the desired local products. Copies of records secured through interactive or batch processing from the utilities, coupled with authority records generated by authorities' processing programs run against standard Library of Congress name authority files, have become common practice for many libraries. In addition, a variety of records have been purchased from several sources and loaded into the local system—again, with standard authorities' processing. Efforts are made to minimize manual review of such data, using, wherever possible, machine-comparison of key access points for accuracy and consistency within the file. Information in local circulation systems has been added to it, matching on key data elements and thereby transferring barcode and item data circulation control information into the online environment. A manager with vision is able to define this new pattern of record creation and transfer, gaining staff support to put it in place. Without it, one is hampered by the labor intensive nature of outdated manual systems.

PLANNING/FORECASTING

Strategic planning is a common process for most libraries in the 1990's. Consideration of the current state of automated applications and the annual changes offered by vendors are essential parts of the annual budgeting and planning process. For a library to move forward its electronic information structure, these must be well defined in the management planning sessions. In turn, the technical services manager must be an effective spokesperson for technical services' needs and present technological applications to colleagues for consideration in the overall library planning process. The successful manager has to couple a solid understanding of technology with a solid knowledge of library operations. These, together with a familiarity with the MARC format and the emerging electronic information publishing patterns, combined with management information and interpersonal skills, create an energized, focused staff who are able to realize organizational goals in this very rapidly changing, dynamic, information service environment.

It soon will be common practice to provide patrons direct, transparent access to multiple institutional catalogs, not just the local catalog. While some libraries already have accomplished this in local consortia, efforts are underway to link series of local systems into one search stream for the patron, providing access to desired sources in a chain of managed databases.

Forecasting, often five years ahead, is an essential part of the annual planning cycle. Without such a forecast, the technical services manager very quickly may realize too late that equipment is outdated or a datafile may not be transferred readily to a new system without major manipulation at considerable expense. Staff may very quickly determine they have lost the ability to process new materials in a timely manner. Critical fiscal resources may need to be diverted to long-term reconfiguration of the hardware and software that serves the online catalog system.

The technical services manager must be able to deal with short-term developments. The unit head also must have the mechanisms in place to keep current with information on developments in hardware, software, telecommunications, market price fluctuations, and mechanisms for accessing local expertise to provide the desired technical information. Networking with counterparts from other institutions can be very beneficial in maintaining currency in newer initiatives to address technical services operations. Without forecasting for succeeding systems needs which rapidly will overtake existing applications, building budget projections, and providing time to incorporate these concerns into the organizational planning process, the unit manager will rapidly fall behind.

Such individuals must be willing to make decisions, often with limited information; to work long hours; to initiate and accept creative innovations, being especially receptive to staff suggestions for new procedures; and to hire staff who are innovative, creative, and, perhaps, more knowledgeable than the manager. The manager must be comfortable in working with such individuals and reward them appropriately for the skills brought to the work force. There is an element of risk-taking involved in the successful management of these units. Such skills are fundamental if the library is to be successful in meeting its objectives of providing up-to-date information resources, especially to the business, medical, and competitive research community of today.

LIBRARY SYSTEMS

Technical services relies heavily on library systems. In many larger organizations, it is not uncommon to find that the technical services manager has assumed responsibility for library systems. One suddenly is faced with a whole new range of administrative responsibilities and demands for business acumen as one attempts to meet with a variety of product vendors, work with staff involved in a significantly different series of operational tasks, and interact successfully with the larger computer support group that many libraries rely upon for their hardware and software implementation. In other cases, the hardware may be the direct responsibility of that technical services manager.

In some settings, the production of media is a part of these responsibilities, as is the preservation of library materials. While microfilm has been a common staple for such activities, this area too is changing as more and more voices are being raised in the request for support for preservation of nonbook materials, especially electronically generated information sources.

On the horizon are more sophisticated workstations to facilitate the cataloging activities and other functions of the staff member, moving from microcomputer applications of some traditional technical functions to the use of scanning and digitizing applications; laser-based technology, such as barcoded item record creation; and a wide range of other components.

Recent changes have produced an environment in which the technical services manager is overseeing a complex operation and relying on automated sources for bibliographic records and the use of mainframe and microcomputer equipment for the production and access to the local online file, either institution-based or regional facility. Faced with meeting the demands of access to an ever-widening group of materials, the individual is challenged to supervise staff who may be well advanced in technological applications but lack awareness of the need for consistency in standardized applications that are essential if the bibliographic information file is to serve the institution into the next century.

Planning here is even more important because the manager must be well versed in the wide range of technical services operations and be able to maximize the skills of each employee to realize the fullest potential. A leader must be able to make frequent changes in the organizational structure, which must be more flexible and dynamic, and to shift resources in a timely manner across units and functions when needed. Public relations for the technical services units will become more important to ensure that patrons and other staff are aware of innovations taking place to meet the needs of particular groups of patrons for specialized materials within the collection. Fund accounting, order placement, preparation of bibliographic records, binding systems, labelling operations, circulation functions, and management information system reporting are aspects of the modern technical services automated system. Unlike the manager of technical services several years ago who tended to be heavily involved in the day-to-day operations, most of today's managers will supervise, coordinate, plan and budget, delegate, prepare management data and analyze it, and learn about new equipment that will enable them to streamline operations. It will be a function of providing direction and leadership rather than actually doing the task of acquiring or processing materials. It will be a heavy complement of balancing the need to provide training for staff and to provide continuing education for the supervisor. This will need to be incorporated not only with training for the unit itself but with providing training for other parts of the library system that need to know about the online data and its structure to enable them to interpret it for patron use.

Presently, it is not unusual to find within a library multiple electronic information sources. While the online catalog is on a mainframe, there are varieties of CD-ROM based-products with their dedicated players, access to remote national information sources through dial-up services, and specialized information files mounted on local information servers.

The availability of full text in libraries is an emerging technology where the contents of a book are made available on a network or minicomputer within the library. One of the challenging roles for the head of technical services in concert with library systems colleagues is providing the network structure for remote access to such diverse files and to make multiple information sources available through as few channels as possible. Front-end menus on larger systems enable the patron to select among the OPAC, electronic datafiles, journal indexes and contents information, and CD-ROM products. Currently, each of these formats is available within the institution, but there are few, if any, reliable mechanisms for bridging access with common communication protocols and search strategies. Libraries have a variety of hardware to support access to these different data-files. Library staff are being called upon to review these several products and determine how best their contents can be made accessible to library patrons in as simplified, straightforward, and user-friendly a manner as possible. How these come together is one of the challenges for the current generation of library managers, including the head of technical services.

The manager of technical services is challenged by technological applications across several divisional operations. This includes a wide range of staff skills in the use of the automated tools now available. Staff need training in the use of various workstations that provide the mechanisms to interact with the online environment with a wide variety of software products that facilitate preparation of materials for patron use. An ever-widening range of formats is received on which information is delivered to and incorporated into the library. New approaches are required to prepare descriptive bibliographic records and compatible equipment to enable the new products to be processed and made usable by the library patrons.

BIBLIOGRAPHY

1. John P. Kotter, "What effective general managers really do," *Harvard Business Review*, Vol. 60, No. 6, November-December 1982, p. 156.

2. Sheila S. Intner, "Copy Cataloging and the Perfect Record Mentality," *Technicalities*, Vol. 10, No. 7, July 1990, p. 12.

Managing Technical Services in the 1990's:
The Ruminations of a Library Director

Delmus E. Williams

As one begins a new decade, it is important to consider where library operations have come from in recent years, the factors that have shaped them, and the factors that are likely to influence their future. This is particularly true in technical services units. Technical services operations are emerging from a period of tremendous change based on the introduction of new technologies, and those who manage them have been forced to adapt to an environment that is far different from the one that they anticipated when they entered librarianship. But it is the premise of this paper that change has only just begun and that the changes in technical services operations that will take place in the next few years may be more fundamental than those that preceded them. While we have introduced new devices into libraries to assist with the organization of information and the delivery of that information to patrons, we are only now beginning to look at how best to organize ourselves to insure that all of our resources can be used to best advantage.

It is important to note at the beginning of this discussion that the mission of the library has not changed in recent years although the interpretation of that mission has broadened significantly. Libraries are still expected to provide to their clients information, books and related materials to support instruction, research, personal enrichment and leisure activities. This involves the development of appropriate collections, the creation of a set of bibliographic tools to provide access to those collections, and the development of a service

Delmus E. Williams is Director of the Library at The University of Alabama in Huntsville, AL.

program that can facilitate the use of an increasingly complex array of materials. The activities of the library are generally divided between those that deal directly with the clientele of the library and those, like technical services, that provide the tools needed to support the library program.

While technical services operations have traditionally been tasked with acquiring materials, accounting for them, and creating the catalogs that serve as the primary finding tools for the library, it is becoming increasingly difficult to view the people who carry out these functions without looking at the library as a whole. Technology has offered libraries opportunities to tailor support operations to meet specific local needs and has increased productivity in technical services while budget constraints and new demands for services have created additional needs elsewhere in the library. If a manager expects to address these service requirements and use all of the resources available to the library well, close cooperation is required among all of the departments of the library. As a result, it is becoming less helpful to view support services outside of the context of the overall service program of the library.

1990 finds us at an interesting point in the history of libraries in general and technical services units in particular. The problems that libraries are likely to face have been defined for the foreseeable future, and the assets available to cope with those problems have been identified. It is unlikely that most libraries will enter a period of dramatic growth or decline or that traditional mechanisms for delivering information will be phased out. We may add new formats for information delivery, but the nature of these is predictable, and it is likely that they will supplement existing formats rather than replace them. And, while the tools that we will be working with have been changing dramatically, it is now possible to identify those that will be used to buy and catalog books for some time to come. For at least the next ten years, technical operations will be defined by things like the MARC format, standard communications protocols, and computer protocols that are either currently in place or that will be compatible with those now in use. While the technology will undoubtedly improve over time and will become more widely available, librarians now know what computers and other technologies can do and, in many cases, have had experience using them. For the first time in a number of years, it is possible to say with some confidence that technical services librarians can now identify those elements within their organizations and the environment in which those organizations exist that will determine the directions that are likely to follow for the next ten years.

Over the next decade, the challenge for technical services managers will be to organize their activities as efficiently as possible to meet the demands of library operations. Their capacity to meet this challenge will be defined by four factors—namely, technology, budgets, the library public service program, and the personnel resources available to the library and to the technical services operation.

TECHNOLOGY AND TECHNICAL SERVICES

At this point, the easiest of these factors to understand is technology. It has been a lifetime since Ellsworth Mason's argument that computers were inappropriate for libraries (published in 1971 in an article called "The Great Gas Bubble Prick'd") was required reading for library school students. The debate about the applicability of information handling technologies has moved to a stage in which the questions used to address new ideas are how the technologies will be applied, how soon applications software will be available, and how libraries will pay for them, rather than whether there will be a need for such applications. Computers have become a mainstay in library operations. Due to widespread use of bibliographic utilities and the widespread adoption of both large and small computers, librarians have moved from viewing the computer as an interesting oddity to a point where even small libraries consider the purchase of integrated automated systems. Discussions of complex telecommunications networks have replaced debates on the importance of computer literacy.

Technical services operations were the first to demonstrate the efficacy of automation for library operations. As a result, those who work in the support areas have gained access to an array of user-friendly programs to help identify materials for purchase, buy materials, check in periodicals, catalog books, maintain accounting records, create databases for online systems, and access those databases as libraries have moved to newer technologies. For libraries that cannot afford a mainframe based integrated library system, there are now less expensive systems designed for medium-sized libraries that are very good, and, for even smaller libraries, there are good systems coming into the marketplace that mount on microcomputers or make use of CD-ROM technology. It must be assumed that these systems will become better and cheaper as more powerful small computers become available. At the same time, open systems protocols and local area networks are redefining the issues that relate to the location of the library computer and the selection of the mainframe. Existing E-mail networks and other affordable methods which offer alternatives for transferring information among colleagues, from library to vendor, and from library to library have also had a significant impact on how libraries operate, and there is every indication that use of these channels will increase. The bottom line is that even the most ardent technophobes in libraries have come to appreciate tools ranging from OCLC to BIP Plus to Bibliofile. As Paul Dumont (1988) put it, the terminal has become the "basic tool of all technical services librarians as once was the National Union Catalog" (p. 59), and this dependence will increase.

For technical services librarians, the only reasonable conclusion is that appropriate technologies will continue to emerge with applications that can improve productivity. While it is clearly possible that whole new technologies may appear, it is more likely that those already in existence will become more sophisticated and play a more significant role in library operations. In cataloging, faster, smaller computers tied together in local area networks are likely to provide viable alternatives to existing mainframes, and programs incorporating the advantages of ever faster microchips and some aspects of artificial intelligence will gradually replace existing software. Public access software is likely to become increasingly more powerful, and development will continue on programs that allow systems to incorporate bibliographic records from more and more sources in both MARC and non-MARC formats into a common database. While quality control in OPACS is likely to continue to be an issue, it appears that the increasing speed of processors and the declining cost of storage make these arguments less important in the grand scheme of things.

In acquisitions, it is clear that useful databases will become available in a variety of formats. Over the last decade, the information community has discussed at length new mechanisms for delivering scholarly information to users, but for the foreseeable future these will supplement rather than replace existing technologies. As a result, libraries must continue to seek ways to acquire both print and non-print materials as quickly and as cost effectively as possible from an increasingly complex array of sources and to enhance their capacity to account for purchases. In ten years it is likely that libraries will order most of their materials electronically, that electronic bulletin boards will make out-of-print buying a more reasonable activity for all libraries, that gifts and exchange lists will be communicated over phone lines, and that invoices for periodicals and records of accounts will be sent electronically as a matter of routine. Most of these things can now be done, and perhaps the most likely change is that these capabilities will become more flexible and will be used more widely.

Perhaps the most important new technology that is emerging is optical data storage, but even this can best be viewed as a variant of earlier computer applications, since CD-ROM players are peripherals for microcomputers, use formats that emulate those that have long been in use for online searching, and provide access (at present) to databases that are also available in either paper or computerized format. Libraries have made an effort to understand how this will affect their operations, but the implications of CD-ROM, WORM and related technologies are not year clear. Some view these products as a significant part of the information community's effort to preserve and disseminate information, while others feel that the declining cost of electronic storage will likely make CD-ROM the microcards of the 1990's. But it is more likely that optical storage will be neither a panacea nor something that will pass out of our lives leaving nothing but outmoded equipment. Instead, this technology is likely to take its place between electronic storage and microfilm and fill a specific niche within the information industry.

But the biggest changes that libraries must face regarding technology have less to do with how the technologies will develop than with how we implement programs to make effective use of those technologies, how these technologies affect our staff, where we get them, and how we pay for them. Programs that allot one to two years in planning for the adoption of technologies that have a life span of five years will not work. Libraries must develop mechanisms to continually identify useful technologies, develop pools of capital to allow for timely investment in those technologies, and develop amortization programs and attitudes to allow for the upgrade or replacement of technologies as advances are made. Investments in technology must be seen as alternatives to investments in personnel costs, and libraries must learn that regular investments in equipment are the only way to insure that the productivity of the library staff keeps pace with the demands being placed on it. Risks must be taken if technology is to pay off for libraries. The costs associated with taking risks are declining, while the costs of not taking risks are becoming higher both in terms of lost opportunities for users and in terms of lost efficiencies. The costs of risk-taking can be minimized if organizations develop the kind of flexibility that will allow them to use innovations in an appropriate way.

LIBRARY BUDGETS AND TECHNICAL SERVICES

The second factor that is likely to affect libraries and their technical services operations during the next decade is money. The financial problems of libraries are not new, and funding for library programs has never been as plentiful as one might wish. But in a political climate in which governmental funding is tight, in an atmosphere in which administrators in higher education are increasingly reluctant to raise tuition, at a time in which the price of books and journals is rising significantly faster than the cost of inflation, and in a period when investment in expensive technologies has become a cost of doing business, budgeting has become even more difficult.

Thomas Shaughnessy (1982) says that increased costs are causing a crisis or turning point for most libraries, and this cannot help but affect every segment of the library program. Much has been written about the dampening effect that this is having on collection development. But it must also be understood that libraries will build collections, they will invest in technology, and they will make cuts elsewhere if necessary to pay for those investments. Given that the "other place" that costs the most in the library is technical services, the needs of the library are likely to lead to a shift of resources away from that area, resulting in an increase in the demands that will be placed on the remaining support areas of the library. It is easy to see from the director's suite that one-time investments in the production operations in technical services can be offset by increases in productivity. Expectations have been growing regarding the dividends that should accrue from these investments. It is becoming less and less useful to argue that automation improves the quality of the library's datafile and increases the amount of management information that is available when libraries are having difficulty paying for their journal list. Automation can do library jobs as well as the manual system less expensively in many cases and this will become more important to the operation of the library.

PUBLIC SERVICES AND THE TECHNICAL SERVICES OPERATION

The third factor that is likely to have an impact on technical services over the next decade is the development of the public service program of the library. Every time that the library increases accessibility to information, new constituencies are created along with new demands for public service staffing. Access to the OCLC database and the availability of easy-to-use CD-ROM or online indexes have increased interlibrary lending traffic; the availability of OPACS, online databases and remote access to library catalogs have brought with them increasingly sophisticated questions about the organization of library materials and calls for more bibliographic instruction; better collections and curricular reform have increased the demand for traditional reference services; and the list goes on. While some libraries would like to find new funding for any new service that might be offered, no one knows better than a director that funding must always be found for any new program. Any library that wants to introduce new programs must be prepared to absorb some of the costs, at least initially. As a result, managers are continually looking for opportunities to reallocate resources to meet new demands for service, and technical services will continually be asked to provide some of these resources.

STAFFING TECHNICAL SERVICE PROGRAMS

The final and perhaps most difficult factor that will have to be dealt with in technical services in the next few years relates to the personnel who work in these areas of the library, including both librarians and support staff. The people who will lead technical services units into the twenty-first century are working in libraries today, and many of them had their last experience in library education prior to the introduction of computers into the curriculum. It is traditional for technical services librarians to consider themselves neglected and unappreciated, even though this is belied in academic and research libraries by the fact that a lion's share of the staffing budget is spent in these areas. But even though libraries have been willing to invest money in technical services, library directors have continued to view the backlogs there, the demands for new staffing to reduce those backlogs, and the productivity of their workers as a "black hole" into which resources are continually sunk without result.

Technical services librarians are expected to run complex production operations that are highly bureaucratized but they are not always well prepared for their task. Those librarians began their careers by defying the wisdom of many instructors in library schools who viewed technical services jobs as unworthy of a professional's time and energy, and they have spent a good portion of their professional careers listening to colleagues in reference units and elsewhere talk about the higher professional calling of public services. They have had little training in management, except perhaps a single course in library school spent reading pop management books or collections of essays decrying the evils of bureaucracy, and they have little work experience in organizations that are not working diligently to mask their bureaucratic nature so as to appear to be participatory. While the best technical services librarians have had an opportunity to learn their craft under older professionals willing to serve as mentors or have honed their skills through involvement in professional activities, many (including the author) learned their craft as the only librarian in an area and with minimal guidance from others in the organization. Risk-taking has not been encouraged, and failure is recognized more often than success. It is not surprising that many of those who work in technical services are what Pauline Cochrane

(1984) called "ritualists" who overconform to professional norms as a protection against challenge from their peers (p. 52). Systems that rely on on-the-job training and do not encourage risk taking may produce good technicians, but they are not suited for developing people who can find new solutions to problems. Perhaps the remarkable thing is that technical services operations have been as dynamic as they are.

The problem of finding leadership for technical services units has been exacerbated in recent years by trends in library education. Library schools have never spent much energy on preparing librarians to work in serials or acquisitions, and many have watered down cataloging courses by expanding their content to the point that they do not include enough exposure to cataloging practice to allow the student to understand the complexity of bibliographic control. Fewer courses are available relating to cataloging and classification theory and practice, and many of these incorporate more theory than practice and more information about trends in cataloging than about the logic of the rules being applied. At the same time, there are more courses available to divert students away from these "skill" courses. Library schools have not encouraged their best graduates to enter technical services, and the supply of people entering these areas has not been adequate to meet the need. Donald Riggs (1988) suggests that few librarians were drawn to technical services prior to automation, and while he senses a change in this trend, that change is not readily apparent, at least not in smaller libraries.

The above may seem a bit extreme, but it is clear that the preparation given to future leaders in technical services has not been adequate to prepare them for the complex decisions that they are now being asked to make. Librarians are expected to function as specialists who can serve as experts in their areas and as advocates for those who work under them while focusing on the goals of the library, and this is difficult for people who have been prepared as generalists.

It is also important for a manager to note that the contribution of librarians to technical services operations is relatively expensive. While their salaries may not be high, the amount of time that librarians are asked to spend on committee work, on professional activities of various sorts, and, at least in many academic libraries, on activities that relate to faculty status and tenure (e.g., publication) mean that much of their time is spent outside the units in which they work. While this may be valuable in terms of the exchange of information that takes place as a result of their involvement and in terms of professional development, that value must be weighed against the needs of the functional unit when roles are being allotted among librarians and the paraprofessional and clerical staff of the library.

This brings to mind another very important element of the staffing of technical services that has been a traditional source of strength for these operations. Libraries have always relied heavily on nonlibrarians to carry the burden of day-to-day operations, and nowhere has this been of more help than in technical services operations. These people bring a variety of talents that are not taught in library education programs, and these talents have been used to good advantage. While it has long been recognized that paraprofessionals and clerical staff members carry much of the work load in production activities, automation has highlighted this fact. As Gisela Webb (1988) noted, "In many cases, highly skilled and eager paraprofessionals have exhibited more creativity and risk taking in assuming new responsibilities than professionals who have difficulties in letting go of comforting, but routine tasks" (p. 113). Libraries are extremely dependent on their efforts to keep their operations moving.

But the utilization of nonlibrarians has been limited by a caste system within libraries that actively penalizes them in terms of status, pay and benefits, no matter what role they play in the library program, and this is no longer reasonable. Libraries must begin to realize that it may be more appropriate to staff some technical and management positions with nonlibrarians than with librarians, and that people in these positions must be given the credit and financial rewards if they are to accept the kind of responsibilities that we are asking them to accept.

ORGANIZATIONAL STRUCTURES
AND THE FUTURE OF TECHNICAL SERVICES

The challenge for technical services units in the next decade will be to organize themselves to use all of their resources as effectively as possible to meet the requirements of a changing library. The one constant in all of this is that there will be change. As a result, the key to success for managers in technical services will be the degree to which they can increase the flexibility of their organizations within the context of the library as a whole. The library director must be able and willing to work within the existing budget, focus the attention of the staff on the job to be done, exploit emerging technologies to best advantage, and expand service programs to insure that the clientele of the library can use information resources in an appropriate way. Within that context, those who manage technical services will be asked to organize their charges to insure that the cost of acquiring materials and building a bibliographic database are controlled without sacrificing staff morale or reducing the quality of their work below an acceptable level.

While the changes that will take place in library organizations may be more subtle than those of the last few years, they are likely to be more fundamental in that they will address issues like the traditional delineation between public and technical services, definitions of library professionalism, and the roles that librarians have claimed for themselves within organizations. It is likely that these changes will have to be accomplished within the context of a traditional, bureaucratic organization with some provision for a variety of trans-organizational structures (e.g., committees, task forces) designed to enhance communication within the organization. Thomas Shaughnessy (1989) suggested that organizational structure follows technology, but it seems at least as likely that structure only accommodates technology while conforming to a model that is comfortable to the manager and to those who work in the organization. But no matter how one arrives at a particular organizational structure, the measure of its appropriateness should always be the degree to which it succeeds, rather than some more elegant theoretical measure. And there is little doubt that comfortable organizations that allow focus while adjusting to the strengths and weaknesses of those who work within them work well.

Following this line, it is reasonable to assume that many librarians who have been "raised" in traditional organizational structures will find those structures comfortable. Some libraries have begun to change their structures in a way that de-emphasizes the public/technical services dichotomy with some successes. But there is very little evidence that most libraries are willing to discard more traditional structures except perhaps to change the titles assigned to these areas, to provide some job enrichment, or to slightly alter the mix of who reports to whom. More horizontal structures that emphasize more holistic approaches to service have a place in libraries, and they may, in some cases, be more efficient in delivering services to the public, and arguments that such horizontal structures give more flexibility in that they move more librarians and support staff members closer to points at which service can be delivered are interesting. But, in many cases, the skills that make a good technical services person do not always fit people well to deal with the public, and those who make good reference librarians do not always indicate great interest in working at the level of detail required of people in technical services. Splitting a position between reference and cataloging may work, but it is just as likely that it will create a good reference librarian who works less well in cataloging or vice versa. As a result, the degree to which full scale integration can take place will depend on the staff in the library and their interests. Hierarchies that clearly define the tasks to be performed and that outline reporting structures based on specialization are easy to understand and provide a structure that is helpful for some staff members. While they may not be elegant from a theoretical point of view, they are not likely to change so long as the organizational structure that is in place works, nor is there any particular reason that they should change so long as they fit local needs.

In some cases, it may actually be necessary to structure organizations so that technical services can operate independently. As building space becomes more expensive and choices have to be made as to what activities will be located away from the library, it may be useful to move technical services functions to remote locations. Technical services used to need space close to the catalogs of the library, but this is no longer true when any terminal, anywhere, can provide the same kind of access to library collections as those located next to the reference desk. While a physical separation of cataloging or acquisitions might not be ideal from a morale- and team-building perspective, it is no worse from a management standpoint than having branch libraries. As collections grow and as resources become tighter, more thought will have to be given to the degree to which support activities need to be assigned high profile space unless they contribute directly to the delivery of service to those who come to the library.

But even if the structure of the library remains the same, the complexion of the library organization will change dramatically. Whatever else happens, the professional presence in technical services is likely to be diminished, and the roles that librarians play in these departments will change. Over the next few years the number of items coming into library collections is unlikely to increase dramatically, electronic ordering services and databases that help identify and find desired items will become more readily available, and reliance on bibliographic utilities and other labor saving devices will continue to rise. As backlogs are cleared and retrospective conversion projects are concluded, managers will have to take a hard look at the number and kinds of people that they employ in technical services. Librarians are hired because they bring special skills and access to information to the library. As cost becomes a more significant factor in library operations, the hiring of librarians may prove to be a luxury that has to be rationed, particularly if this information can be brought to the library more efficiently.

It is unlikely that most libraries will be willing to do without at least one professional cataloger, even though an increasing amount of the cataloging will be taken from bibliographic utilities. But libraries may find it is easier to instill an understanding of the larger mission of the library into a person who has had years of paraprofessional experience in the library and perhaps has a background as a bookkeeper or a microcomputer specialist than to teach a new professional to understand how the local system works and how to work with the staff at hand. The criteria that we use for deciding who will lead departments are working less and less well. While it may never have been reasonable to reserve jobs for degreed librarians without carefully considering the specific requirements of the jobs and the people who might fill them, it is increasingly clear that the need for different mixes of skills and financial constraints requires managers to revisit this question.

As time progresses, it is important that the role of those who manage technical services (either librarians or nonlibrarians) and other librarians who work in those units changes. As Brian Alley

(1988) notes, the librarians in an organization are likely to move into positions that require them to train, manage, and supervise staff. They will be hired as experts who are expected to use their special skills in roles as leaders, specialists and trainers-in-residence to provide direction to their areas, help create an appropriate set of values in the organization, and insure that the organization exercises appropriate quality assurance. These people will also be expected to provide bridges to innovators outside of the library and serve as conduits through which new ideas can be introduced into the library. They will have to be recognized and compensated for the special skills that they bring to the library and encouraged to continually upgrade their skills in their specialties. The library administration must be able to balance the need for productivity with less tangible qualitative measures to insure that operations function well. Arguments about professional status and definitions of professional rules may be useful, but librarians must understand that professionalism can only be defined in terms of the specific roles people play in the organization in which they work. The kind of values and traits that make a good cataloger or acquisitions librarian must be given standing equal to those that make a good public service professional or administrator when those who hold these positions are hired, considered for retention, promoted or tenured. Just as we now separate the function of the cobbler from that of the shoe salesman with value given to the special skills required to perform each job, so we must be willing to recognize that a good cataloger may need different skills than a good circulation librarian and that a different career path may be required for people holding each of these positions.

The definitions of the roles of those who work in technical services who have not had the benefit of a library school education is likely to provide a real challenge in the coming years. Both Cochrane (1982) and Alley (1988) have commented on the degree to which nonlibrarians have taken over leadership in the adoption of new technologies, and it is generally understood that much of the supervision of day-to-day operations in technical services has passed or is passing to nonlibrarians. But, to date, librarians still maintain the fiction that they are the only "professionals" in the library and that others somehow have a lower status. While this may be true in a narrow sense, the distinction is becoming less helpful as libraries try to maximize the efficiency of their operations. It is important that libraries expect professionalism from all workers, and this requires that caste distinctions be eliminated and that a more personal approach to management be taken.

Changes will also have to be made in the way we look at nonlibrarians when dealing with matters of policy and when considering compensation. Support staff members who learn new skills will have to be paid more, and, if these employees are to be given managerial slots, mechanisms must be developed to allow them to sit with librarians in policy making sessions when it is appropriate and to pay them at levels commensurate with their responsibilities.

It is also likely that the amount of staffing required to perform technical services functions will decrease. But, if one agrees with Dana C. Rooks and Linda Thompson (1988) that "no job should be structured so that a staff member is tied to a terminal all day repeatedly performing a single function" (p. 126), the number of people assigned to those tasks may not change. As a result, individual jobs must be drawn broadly enough to allow the staff members' skills to be used fully. Increasingly, library managers are going to have to look at the array of functions that are required to make both technical services and other areas of the library work, assess the potential of the people working in various positions in the library, and tailor positions to meet the needs of the library while offering the staff satisfying work. This kind of job enrichment will require that the manager be willing to limit the amount of control that he or she has over the operations; show a continuous interest in the details of the tasks to be performed in the unit; use creativity in dealing with individual employees; and provide employees incentives for learning new skills and opportunities for advancement. While this can be accomplished within a traditional hierarchy, it will require some breakdown of the demarcation between areas of responsibilities and more cooperation among middle managers.

The organizational changes coming to technical services are likely to cause substantial discomfort. The challenge for the manager will be to ease the transition and harmonize the values of those who work in these areas. These managers will be forced to gain a more intimate understanding of the tasks that contribute to success in technical services and to place a higher value on flexibility in the organization so that positions can be moved to other areas in the library or eliminated with minimal discomfort. The technical services manager will be expected to function as a floor manager who plans constantly to integrate new technologies, consults widely, understands the strengths of those who work in the unit, and tailors positions to match the strengths of the staff with the needs of the library. Everyone in the library will be expected to understand the technologies in use in the library as the planning and implementation cycle for innovation shortens dramatically. Managing a department in this kind of situation requires a patience and maturity that is difficult to teach and learn, but it must be found if library operations are to be as productive as they need to be.

CONCLUSION

Managing technical services in the next ten years may be more difficult than it has been at any time in recent memory. Evidence that change is taking place will be far more subtle than in past years, and credit for the kinds of changes that will be made will be hard to come by. But changes must be made in basic attitudes toward job assignments, in librarians' views about professional status, and in the way libraries make important decisions, and these will be as significant as any changes made in recent memory. Basic values must be re-examined, and, when they are found to be out of touch with existing realities, they must be changed. This will pose a severe test for those asked to manage technical services programs and should make such programs interesting places to work for the foreseeable future.

NOTES

Brian Alley. (1988) "Reshaping Technical Services for Effective Staff Utilization." *Journal of Library Management.* 10: 105-110.

Paul E. Dumont. (1988) "Creativity, Innovation and Entrepreneurship in Technical Services." *Journal of Library Administration.* 10: 57-89.

Janet Swan Hill. (1988) "Staffing Technical Services in 1995." *Journal of Library Management.* 10: 87-103.

Gillian McCombs. (1986) "Public and Technical Services: Disappearing Barriers." *Wilson Library Bulletin*. 61: 25-28.

Ellsworth Mason. (1971) "Great Gas Bubble Prick'd; or Computers Revealed — By a Gentleman of Quality." *College and Research Libraries*. 32: 183-196.

Donald E. Riggs. (1988) Leadership versus Management in Technical Services." *Journal of Library Management*. 10: 27-40.

Dana C. Rooks and Linda L. Thompson. (1988) "Impact of Automation on Technical Services." *Journal of Library Management*. 10: 121-136.

Thomas W. Shaughnessy. (1989) "The Impact of Rising Costs of Serials and Monographs." *Journal of Library Administration* 11: 3-15.

Thomas W. Shaughnessy. (1982) "Technology and the Structure of Libraries." *Libri*. 32: 149-155.

Gisela M. Webb. (1988) "Educating Librarians and Support Staff for Technical Services." *Journal of Library Management*. 10: 111-120.

Beyond Our Walls:
Academic Libraries, Technical Services and the Information World

Arnold Hirshon

We need to see more of the big things in the little things; it reinforces the relationship of the whole to the part.

— Frank Lloyd Wright[1]

THE INTERNAL AND EXTERNAL WORLD VIEWS

Academic libraries today are complex multimillion dollar organizations that are often little understood by their constituencies. Often this lack of understanding is caused by a lack of adequate explanation by the library for the actions it takes. For example, most libraries replaced (or are replacing) their card catalogs with online catalogs. Yet there are still some faculty members who do not understand why the library no longer has a card catalog. It is clear that the library failed to explain the expense of maintaining a card-based system or the improved services available through an online system.

An effective analogy to explain how the library relates to the rest of the information world is to compare the two classic astronomic views of the universe. Using a Ptolomeic (or "internal") view of the information world, the smallest organizational unit (such as technical services) is put at the center (See Figure 1).

While it is reassuring to library staff to think that the information

---

Arnold Hirshon is University Librarian at Wright State University, Dayton, OH.

FIGURE 1

PTOLOMEIC VIEW

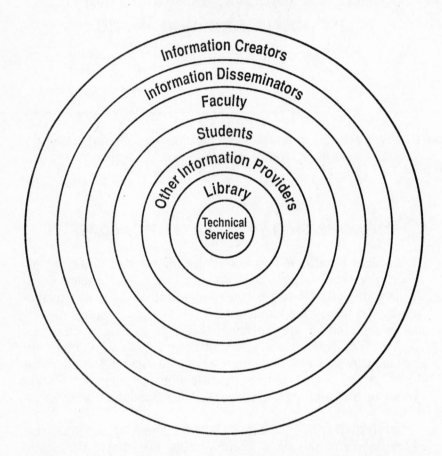

world revolves around library technical services, nonetheless it is dangerous to move outward from the specific to the general. Technical services is not the center of the library world, nor are libraries the center of the information world. If nothing can occur at the outer edge of the information world without revolving around some action library technical services takes then libraries would dictate developments in the information world rather than respond to those developments.

By contrast, a Copernican (or "external") world view puts the information creators at the center of the information universe (see Figure 2).

In this view, technical services is one of many information organizers and providers. The library exists in one of the outer orbits. This view illustrates that libraries and their operations not only work with the general public, university faculty, and administra-

FIGURE 2

COPERNICAN VIEW

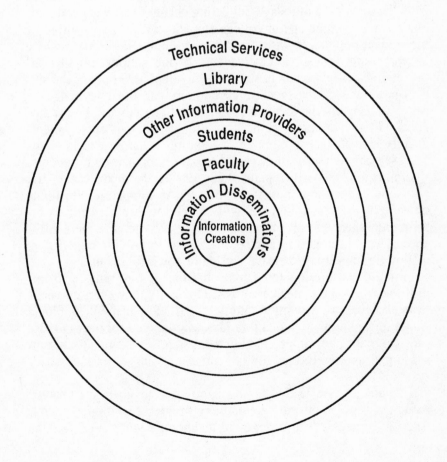

tors, but also with members of the Friends of the Library and with donors, publishers, and legislators.

To respond to the needs of its constituencies, the library should harmonize its mission, goals and priorities with that of its the university. To build an effective externally-based model, however, library administrators must have a strong understanding of the general factors that will affect universities by the beginning of the 21st century.

THE UNIVERSITY OF THE 21ST CENTURY:
SOME PREMISES

The university of the 21st century will be significantly different from the university of today. Reflecting changes in the general population, the average age of college students will be older. More of the student population will live in metropolitan areas. As minority groups compose a greater percentage of the national population, universities will continue to stress multicultural diversity of the student body and on the faculty. Technology and lifestyles will also cause a change in the methods for information delivery. Fewer students will attend classes in the "traditional" campus environment. Faculty increasingly will request re-training to use new technological research methods so they can keep pace with their own students.

Library budgets will remain stable during the next decade, but with a net reduction in purchasing power. As competition for funds on campus increases, libraries will contend with other campus information agencies, all of which will seek to increase their influence. This will create a demand on campus to provide more information through less expensive and more powerful means.

Initially, the driving force for using technology will be to improve services and to decrease costs. Gradually, however, universities will develop a new information infrastructure that will employ a single information system. Librarians who labored for over two decades to realize an "integrated library system" will find this system subsumed as merely one component in an "integrated university system."

Technology is already making it possible to go beyond keyed input and flat text information delivery to interactive digital sound, data, and images. The changes in technology will also affect the scholarly publishing environment. Not only will there be increased interest by faculty and administrators to find ways to reduce the costs of scholarly publications, the capabilities of the new technologies will also make entirely new and more powerful forms of communication possible.

While the full impact of electronic communication will not be seen immediately, there will be significant changes within the next ten to twenty years. As a result of the work of such groups as the Coalition for Networked Information and the development of the National Research and Educational Network (NREN), there will be large-scale national delivery of electronic information. Eventually the predominant form of information delivery will be electronic.

As the information world changes library staff will be under increased pressure to perform and to be knowledgeable in a wide variety of areas. The hiring of individuals who possess abilities will be more important than hiring individuals with skills. Finding qualified individuals will be increasingly difficult, particularly for entry level positions. According to the *New York Times*, there will be a continuing demand for "well-educated entry-level workers capable of making independent decisions . . . increasing the pressure on schools to educate better and on supervisors to change . . . to a [management style] that encourages teamwork and initiative." The *Times* also reports that "Information technology . . . must now be an integral part of corporate strategy . . . providing more opportunity for the computer-literate."[2]

Technology will also change the nature of university communications. Computers with increased speed and decreased size will be powerful, portable, and readable personal information systems capable of communicating from anywhere. Instructors and students will be able to interact electronically, thus making campus residency or physical class attendance less common.[3]

Finally, there will be less literacy in society in general, and in undergraduate students in particular. As evidence, witness the continual decline in the national average scores on the Scholastic Aptitude Test (particularly the decline in verbal scores). Educators may continue to deplore this trend, but the reality is that the new students will seek a world that is more graphically oriented and less text oriented. This will drastically change the instructional role that libraries will have to play.

CREATING A VISION AND FOCUS FOR ACTION

Armed with a clearer understanding of the factors in the outside world that are likely to affect libraries, it is then incumbent upon the library administrator to set forth a *vision* that defines the goals of the organization. There must also be a plan, or *focus*, to define the specific actions that will be necessary to achieve these goals.[4]

An organizational vision goes beyond immediate or parochial concerns to project an image of what the library should accomplish over a period of five or more years. Strong visions demonstrate creativity, perspective, risk taking, and realism.

Organizational focus lends direction for accomplishing the vision, by providing the enterprise with practical steps to bring about change in specific and evolutionary ways. The characteristics of focus include organization, weighing the alternatives, revision based upon experience, recognition of political realities, and patience.

DEVELOPMENT OF A PLANNING MATRIX

The simple and effective planning matrix in Figure 3 combines the concepts of the "internal and external views" of the organization discussed earlier with the concepts of "vision and focus." An administrator can use this matrix to test whether a new organizational vision will be based upon the needs of its users, and whether the plans for executing the vision are sound.

Across the horizontal (X) axis are the factors of vision and focus. Down the vertical (Y) axis are the two ways of viewing the future: the internal (how the staff views the issue) or externally (how the library users would view the issue). In each of the four cells of the matrix are decision points for the administrator. For a particular planning area (such as cataloging or automation) the administrator would set forth both the desired goals (or vision) and specific actions (or focus) as viewed by the staff (internal) and library users (external).

The remainder of this article will illustrate the value of the matrix by using as examples three specific planning areas or topics: the library organization chart; collections and access; and bibliographic

FIGURE 3. Organizational Planning Matrix

	VISION	FOCUS
INTERNAL VIEW [How the staff views the library]		
EXTERNAL VIEW [How users view the library]		

system design. For each of these topics I will set forth a vision and a focus, giving for each the internal and the external views.

THE ORGANIZATION CHART

Internal View. The typical technical services organization chart from the early to mid-1970's looked like the chart shown in Figure 4.

Of course at the time there were variations, such as whether to have serials or bibliographic searching in a separate department, but there would probably be general agreement that this chart was typical.

Compare this to the typical technical services organization chart of today (see Figure 5).

Note the striking differences. Of course there are variations, such as whether to have serials or bibliographic searching in a separate department, but there is probably general agreement that this chart is typical. Which raises the question: why have library organization charts changed so little over the past fifteen to twenty years?

External View. The reason library organization charts are so traditional is that librarians continue to look inside for their inspiration. James R. Houghton, chairman of Corning Glass, says that in business the traditional "hierarchy, with tasks strictly defined by organization charts and tightly drawn job descriptions" were "more appropriate to the 19th century than the intense competitive environment of the 20th." Instead, he suggests that businesses become a closely linked "global network." He goes on to say that "Alliances or joint ventures are a major component of our structure. . . . [and they are] likely to become more commonplace. . . . [because a] partner may add expertise. . . . " This cooperative networking arrangement with external organizations adds "the necessary flexibility and strength to prosper in the future."[5]

Vision. Libraries must abandon the conventional library organization chart, which neatly divides public and technical services. That model is inadequate in an interdisciplinary and interorganizational information world. Library managers should adapt organization charts to a service model that views the library as its users see it. A new, externally-based library organization chart should:

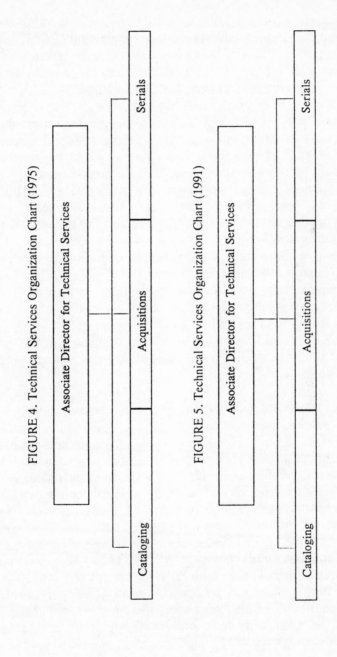

FIGURE 4. Technical Services Organization Chart (1975)

FIGURE 5. Technical Services Organization Chart (1991)

1. reflect alliances with other information providers on campus and outside the university. The organization chart should incorporate collaborative and contractual arrangements with intra- and extra-university departments and organizations.
2. view the library from the outside-looking-in, stressing direct services to the user rather than internal operating procedures.
3. have as flat a hierarchy as possible, with few upper administrators and more direct contact between staff and administration. A flatter organizational pattern with managers close to their operations will increase the information flow.
4. decentralize processing operations to their related public service. For example, the library might make the same staff responsible for both serials check-in and the periodicals information desk.
5. be dynamic, with periodic fine-tuning annually and major overhauls every 3-5 years.
6. employ technology to make committee work more effective (such as greater use of electronic mail to revise documents, with less reliance on face-to-face meetings).
7. reflect a new level of sophistication in integrating technology into the organization, rather than viewing technology as an end unto itself.

Focus. To evolve into externally-oriented organizations libraries should involve users to redefine mission, goal and service statements every three-to-five years. Redefinition in less than three years could lead to a lack of stability of the vision, and the focus could become diffuse as the library chases every new idea. Waiting more than five years could lead to the vision becoming stale. The organization might not be as responsive to advances in technology or differences in the economic climate.

The first task for the administrator is to define which services are critical functions of the library, and which tasks should be obtained or provided from outside sources. Next should be a review of the organization chart to ensure a synchronization between the organization chart and the changes in technology, staff, and the budget. Finally, administrators must provide strong formal and informal communication channels through meetings, newsletters, and policy statements so all staff and library users are aware of changes to the organization and services.

Before the traditional organization chart gives way to the new externally based organizations, most libraries will likely go through two or three evolutionary models. Library administrators should carefully examine what trends will have the most effect on their local applications, and plan their organizations accordingly.

STAFFING: JOB DESIGN, STAFF SELECTION AND DEVELOPMENT

Internal View. Job descriptions often define qualifications for positions based on skills rather than on abilities. This has been true not only at the paraprofessional and clerical ranks, where such skill requirements are appropriate, but also at the professional level. The typical job description in technical services specifies skills such as working experience with AACR2, OCLC, local automated systems, the MARC format, vendor files, or serial check-in systems. In an effort to ensure that candidates possess such skills, librarians sometimes craft interview questions that resemble quizzes in cataloging or acquisitions practices, rather than trying to discern the general problem solving abilities of a candidate.

External View. The external orientation reveals that librarians will need to have a more diverse background to get a job. Librarians will have to work well with people outside the library, be multi-faceted, and be tolerant in ambiguous situations. Critical to success will be the ability to understand not just one's own job specialization, but how that job fits into the larger context.

A precursor to this occurred during the introduction of OCLC in the mid-1970s. The inability to adjust to the new cataloging methods encouraged the retirement of some people who were intellectually tied to a specific methodology (card catalogs). These individuals were unwilling to further develop their abilities in information access. Staff who today are bound to a specific technology (such as the online catalog, a serials check-in system, or the MARC format), are likely to contribute to their own extinction in the same manner as did their predecessors.

Vision: Libraries will be known on campus and nationally by the quality of their staff. The process for hiring and retaining professionals will emphasize general abilities, such as problem solving. To enrich the experience base of the staff and to provide an inviting atmosphere for students and faculty, library administrators will stress the importance of multicultural diversity. In recruitment, libraries will need staff members who work well with teaching faculty and students. Professional staff will need to demonstrate scholarship, leadership, a managerial or subject specialty, the ability to communicate persuasively orally and in writing, and possess an understanding of the academic environment. Staff will work effectively as a team to provide high quality services. Libraries will need well-organized and effective staff training, professional development, and evaluation programs.

Professional advancement will be synonymous with evidence of professional growth. For librarians to achieve the needed recognition from the outside world, library administrators should expect their librarians to demonstrate external professional activities, such as publication and involvement in professional organizations.

The library will become a cooperative endeavor that fosters a good understanding of the importance of all staff contributions to the organization. The library will develop a well-organized and effective program for staff training, professional development, and evaluation.

Technical services librarians will seek the opinions of library users and how technical services might meet those needs. As the role of librarians increasingly is one of information management, there will be a call to maintain full-text databases and to establish new retrieval protocols to access those databases.

Focus: Through a strong recruitment program, library administrators will identify candidates with strong abilities. Staff development will also be a high organizational priority. Training will address both immediate and long-term needs. In addition, the description of positions, particularly at the professional level, will provide for significant responsibility and autonomy. A mentoring program will ensure that new staff are aware of the needs of the organization. The evaluation system will also provide incentives, resources, and opportunities for professional involvement, including formal recognition and rewards for both on-the-job contributions and external activities.

Finally, there will be an expectation that every administrator is acting in a manner that provides a good role model for the staff.

COLLECTIONS AND ACCESS:
PUBLISHING AND COMMUNICATIONS ENVIRONMENT

Internal View. The traditional bent of library collection building stresses local ownership of as many materials as possible. Materials selection concentrates on extant resources in the library. Cooperative collection development does not fulfill its promise because of inadequate availability of information about the holdings of other libraries. In some of the worst cases significant redundancy exists among the collections of the main academic library and those of the institution's separately administered medical or law libraries. Access is less important than ownership. The library creates its major form of bibliographic access, the library catalog, and is highly possessive about the data and concerned about its quality.

External View. Both library budgets and the information world are rapidly changing. University officials require greater accountability for library expenditures. Within the information world, new methods for access are changing the view of the collection. It is now possible to access the on order and catalog information of other institutions online, thus making cooperative acquisitions more realistic. New methods of document delivery, based upon telefax or its successors, will greatly expand off-campus access to library materials.

Vision. The decreased purchasing power of the library materials budget will occur simultaneously with the growth of electronic publishing. This will place contradictory demands on the library to explore new technologies while meeting current demands to purchase traditional books and journals. Unless libraries continue to collect the "critical mass" of print materials necessary to meet curricular and research needs it will be difficult for a university to maintain its institutional identity.

Nonetheless, a paradigm shift is evolving. During the 1990s the "library materials" budget will gradually become an "information resources access budget." Librarians will develop new methods for allocating budgets, and demonstrate increasing concern for preserving digital and non-print text that no longer has market value. These changes will require not only new budget allocation systems, but also new measures of library service.

Focus. To ensure accountability library managers will develop local strategies to ensure that libraries procure vendor services fairly and through open competition. For the 1990s, librarians will create collection formulas to define the "critical mass" of materials to be owned versus those to which the library provides access. As part of a complete collection policy for the next 5-10 years, librarians will develop a desired collection size and expenditure target for each subject area. Library administrators will move away from the concept of an "acquisitions budget" and replace it with an "information resources access budget." The latter will include costs for online databases, CD-ROM subscriptions, interlibrary loan expenses, telecommunication charges, local database purchase, and data conversion. As access increases, the library will need to seek support to expand revenue sources.

The library community will begin to discard the old quantitative measures (such as the number of volumes in the library or the number of serial subscriptions) and to develop new quantitative and qualitative measures to determine what constitutes effective library service.[6] To develop new measures of what constitutes effective library service, libraries should begin by changing the statistics included in local annual reports, and call for changes in national statistical reporting (such as HEGIS, ARL, ACRL, etc.). For example, rather than measure the number of volumes in the collection, libraries should measure the success rate for locating materials both within and outside the local collection.

BIBLIOGRAPHIC SYSTEM DESIGN

Internal View. The design of many of today's automated library systems reflects its origins as a way to solve internal problems such as circulation control, ordering, or the generation of cataloging records. As the information becomes available through online catalogs, there is a sudden realization that much of the information is inadequate or incorrect. Systems are filled with jargon, and are often so complex that the user cannot understand how to use the technology.

External View. The primary purpose of the library system is to provide public access. Background functions should enhance public access. Systems should be simple to operate and provide a wealth of information from terminals that are dispersed geographically.

Vision. A collection is only as good as the systems to access it. As the collection dollar decreases, libraries should provide increased access to that which the library *does* own or can easily obtain. The challenge of the 1990s will be to balance spending for access and networking technology against the continued need to purchase information in traditional formats. Three trends will occur during the next ten years.

First, libraries will expand the usefulness of the local online catalog by enhancing the amount of information in catalog records and by integrating external databases (such as periodical indexes, user guides, and online encyclopedias) for use by multiple users at remote sites.

Gradually libraries will link their separate institutional library catalogs into regional and national networks. Local networks will allow faculty to make electronic requests, such as to borrow items on interlibrary loan, or to request document delivery. A national linkage of local catalogs will provide each professor with a menu of local institutional catalogs to search. Through the use of a common command language, users will search other library catalogs as if the catalog were at their home institution.

The third trend will occur after the year 2000. The expanded library catalogs will decrease in usefulness as new automated publication formats, primarily full-text retrieval, become prevalent. This will cause a need for a major overhaul of current library access methods. Fortunately for users, by employing user studies and application of artificial intelligence, the replacement systems will be more responsive and less prescriptive than the current catalogs.

Focus: Libraries should purchase and install library systems that are expandable, and whose vendor has a demonstrable record for introducing nontraditional access functions. Vendors should demonstrate a strong commitment to academic libraries, and should employ an open systems architecture that fosters networking.

Libraries also should expand the concept of the library information database to provide improved access to the existing collections. For example, the system should provide direct user access through the local catalog to such databases as government documents, MEDLINE, ERIC, NTIS, Wilson databases, ABI/Inform, UnCover, Current Contents, etc.

It is already important to link the library to the campus, state, and regional networks. An additional enhancement will provide an interactive electronic request system for faculty and students to renew books, place items on reserve, ask a reference question, or request document delivery.

Finally, to build for the future, libraries should cooperate with extant projects to expand retrospective conversion of full texts into digitized form.

CONCLUSION

The library of the future will tailor itself to the external environment. Its staff will be rich in abilities and will be culturally diverse. The organization will be understandable intuitively by library users. Libraries will move rapidly toward an electronic system of scholarly communication and organization, and will work in partnership with other faculty and information providers on campus.

Perhaps librarians have not looked more to the outside for direction because the external environment was too threatening, or perhaps because librarians have become too complacent. This is a position that library administrators can no longer support. Technical services librarians must recognize a new service imperative. To survive there must be a change to the stereotypical view of technical services in which the "technical" overshadows the "services."

These new service strategies will affect personnel placed in the positions requiring public contact. There will be a need to cultivate new abilities if libraries are to better serve the public. Staff must be able to both properly process materials and provide clear information to the public. Administrators must be not only superb at analyzing operations, but also effective problem solvers who can be courteous and helpful to library users.

The most important challenge is to improve public relations, and not just superficially. The online system makes apparent to every user a host of problems that were previously buried in the shelflist or some other file hidden in the backroom. With public access systems, users will begin to ask questions; technical services librarians need to respond openly, forthrightly and constructively.

The planning matrix presented here can be simple and effective both for isolating the key issues facing libraries in the future and for developing solutions to the problems raised. To be relevant in the new information universe, librarians must begin to "reduce the whole to its parts" and to "see more of the big things" of the outside information world "in the little things" technical services librarians do on the inside so we can "reinforce the relationship of the whole" to our part.

ENDNOTES

1. Langdon, Philip. "In the Wright Tradition." *The Atlantic Monthly* (April 1989): 87. [Quotation of Frank Lloyd Wright as remembered by E. Fay Jones. Jones was likely referring to the quote by Wright to:

> think in "simples" . . . reduce the whole to its parts in simplest terms, getting back to first principles. Do this in order to proceed from general to particulars and never confuse or confound them or yourself be confounded by them. Frank Lloyd Wright, "To the Young Man in Architecture," in "Two Lectures in Architecture, 1931," in *Frank Lloyd Wright: Writings and Buildings* (New York: New American Library, 1960), 250.]

2. Chira, Susan. "In 1990's, What Price Scarce Labor?" *New York Times*, Sunday, 1 Oct. 1989, Careers section: 29.

3. For a particularly interesting view of the potential effect of technology to change the physical location of instruction, see: Luke, T. et al. "TABLET: Academic Computing in the Year 2000," *Academic Computing* (May/June 1988): 7-65.

4. Hirshon, Arnold. "Vision, Focus, and Technology in Academic Research Libraries: 1971-2001," *Advances in Library Automation and Networking*, v. 2 (Greenwich, CT: JAI Press, 1988), 215-257.

5. James R. Houghton, "The Age of the Hierarchy is Over," *New York Times*, Sunday, 24 Sept. 1989, Business section: 3.

6. Richard Dougherty, "New Measures; or, Bigger Won't Always Be Better," *Journal of Academic Librarianship* 16 (March 1990): 3.

New Partnerships Within the Library

Sharon L. Walbridge

INTRODUCTION

A common theme heard in presentations about the future of libraries these days is the need to build new partnerships or coalitions. These new partnerships are encouraged for libraries and the faculty, libraries and computing centers, universities and business, universities and government, etc. However, it might be well to look inward and begin to consider new partnerships within the library. More to the point, it might be the time to propose what for many libraries would be a whole new way of looking at the organization — as a team with librarians taking a holistic view of their roles. The holistic approach is defined here as every librarian committed to the library's service mission and actively involved in making the library as effective as possible.

There are a number of factors driving the need to reconsider the organization of the library and to encourage new working partnerships among what traditionally have been separate departments. These factors include the impact of the online public access catalog, the inability of the traditional technical service areas to deal with the influx of materials, the ramifications of new technology and new forms of information and increasing competition in the provision of information. This article will focus on the first factor.

The changes proposed in this article will likely effect technical service librarians more than their colleagues for reasons which will be elaborated. However, the changes proposed in technical services may yield unexpected benefits for staff in those areas. These bene-

Sharon L. Walbridge is Access Services Librarian at the Oregon State Library, Salem, OR.

fits will be in addition to those that accrue to the library as an institution.

CHANGES IN LIBRARIES WITH OPACs

The online public access catalog (OPAC) presents librarians with opportunities to better serve the library user through more effective use of library staff time and knowledge. The OPAC is creating the need for new patterns of library organization and opportunities for professional librarians to enhance their role in serving the information seeker. These factors could change the way society views the library.

It has always been theoretically true that the various parts of the library all work together to support the library mission. However, the behind-the-scenes operations of the library have not always been perceived by library users and by public service librarians as sharing actively in that support. The OPAC has served to make the key to using the library—bibliographic information—more visible and vital. Where use of the traditional card catalog seemed to be a private activity, use of the computer terminal seems to bring use into a more public realm. Even the terminology used to describe the newer form of the catalog—the "public" catalog is recognition of that perception. (The phrase "public catalog" was used in some libraries to describe the card catalog, but more often the traditional terminology was simply "card catalog.")

PROBLEMS/OPPORTUNITIES

The online public access catalog has made the catalog more visible in its power—and its deficiencies. Search results seem more obvious and thus the policies, rules and decisions made by the cataloging department are more obvious. Users are more apt to want to print out the results, not just jot down a few call numbers on a piece of scrap paper. (They are getting accustomed to printing results from CD ROM databases and may be receiving printed results from online searching.) When they ask for help in using the OPAC, the screen image or printed list of search results gives the librarian the opportunity to see how effectively the user has searched and also to see the deficiencies of the catalog (index structure, inconsistencies of data, interface difficulties, etc.).

Online users seem more apt to ask for help—it is as though it is more acceptable to ask for help to use a computer whereas everyone was supposed to be able to use the card catalog. And, more libraries are actually providing proactive staff support for the OPACs. This support can be in the form of library staff who approach users who appear to be having problems using the OPAC or a desk or office close to the central cluster of terminals where users can get immediate help.

Many libraries report that online access is bringing heavier demand for library materials and for service in terms of helping users negotiate the system as well as knowledge of and maintenance of the equipment (terminals, printers, etc.). Such increased demand on the public service staff offers an opportunity to evaluate the service role of others on the library staff, specifically technical service staff. The question is, have online services changed the role of the librarian sufficiently to encourage examination of the traditional roles and relations? Can this serve as an incentive to bring those aspects of librarianship (especially technical services) which have heretofore been misunderstood and undervalued into a more visible arena and at the same time to make better use of their knowledge and experience?

Libraries have usually been highly structured organizations whose various units comprised the working functions that made the library operate. However, the people in the various units often did not relate well to one another. The technical service areas did the background work upon which the library was built. The public service areas used the products of technical services—the materials that were purchased by acquisitions, the catalog that was the product of the catalogers, etc. The technical services people seemingly went about their work routines with little thought of the people who would ultimately use the products of their work, whether they were library patrons or public service librarians. Catalogers had their rule books, subject thesauri, etc. They were concerned with consistency and accuracy. While those concerns were not invalid, those who subscribed to them did not necessarily do so for the sake of library users.

With the card catalog, librarians were relatively invisible—catalogers were behind closed doors and reference librarians were not always situated near the card catalog area. Users who did not find what they wanted in the card catalog often just left the library. Now, with public access terminals scattered around the library, librarians are more apt to be around to be asked for help. Librarians also seem to have become more visible with the advent of online database searching—they seem more approachable. In many cases, the existence of online search service is better promoted than other library services have been in the past and this also may affect the way users feel about OPAC support.

For the reference librarian who may not understand the principles of AACR2, may not "speak MARC," and may not understand LC or NLM subject headings, but does see the frustration and confusion of people trying to gain bibliographic information, the dilemma is clear. If the reference staff does not understand the catalog, how can students or townspeople ever use it effectively? But have there been serious attempts to reconcile the problems with dialogue within the library? Perhaps in some libraries, but such attempts were fairly rare. And the problems seem to be more significant with the more visible OPAC.

It is clear that those who build the OPAC—the catalogers—and those who interpret it—the reference staff—would benefit by increased dialogue. No longer can the catalogers make decisions and do their work in an isolated fashion. Even in the days of the card catalog, decisions that effected the use of the catalog should have been made in concert with the public service staff or at least communicated to them. Unfortunately that often did not happen. Catalogers believed that cataloging was their purview. As long as they followed cataloging rules, assigned classification numbers and subject headings as best they could and did authority control, they were doing what they should do—what they had been taught to do. (Do cataloging instructors ever suggest that catalogers in a real library situation should be concerned with public use of their product?) With the increasing emphasis on productivity, catalogers were doing well to avoid building backlogs of uncataloged books. How the product of their work was used was not always a primary concern.

HOLISTIC LIBRARIANS AS A SOLUTION

Simply defined, the holistic librarian is one who keeps a broader view of the library—who keeps the library's mission paramount in mind as problems arise and decisions are made. The holistic librarian is also likely to perform a broader range of tasks. Whether one's responsibilities are reference, ILL, cataloging, acquisitions, personnel, or whatever, the overall good of the library and its users are the first concern. In a sense, this would appear to be a motherhood and apple pie statement, but it is clear that the specific job is the primary concern among most librarians.

The answer is to create an environment that brings the catalogers out into the public to both learn about how the public uses their product and perhaps even to have the catalogers help teach the users how to best use that product. Combine the catalogers' knowledge with the reference online searchers' knowledge to make decisions to make the OPAC the powerful tool that it can be. There must be mutual respect in that dialogue—each party must be committed to making the library's catalog as effective and usable as possible, drawing from the knowledge of cataloging and online searching.

The dialogue between public service and technical service staffs should be regular and ongoing. Problems should be library problems, not reference, cataloging, ILL, acquisitions, etc. Regular meetings should be set to do mutual problem identification and solution. If everyone understands the problems of the library and everyone has the opportunity to contribute to solutions, the result can be a commitment to the library mission by all members of the library staff. The days of catalogers alone considering cataloging problems and setting priorities, for example, should be behind. A recent article by Ilene Rockman in *Library Journal* addressed this topic in respect to retrospective conversion decisions.[1] It should be clear in the OPAC environment that everyone who helps people use the library is aware of problems and deficiencies and should have a hand in solving them. Catalogers have expertise but so do online searchers. It is time to bring them together on equal footing to bring a more effective basic tool to library service.

There are a number of ways of accomplishing this new environment. It can be done with team building, with cross training, with abolition or modification of traditional structures which have created walls between the working units of the library. It can be done by persuading each librarian to take a holistic view of his/her role — being more concerned with the overall view of the library rather than being dedicated to the specific role one plays. All of these methods call for belief on the part of the participants that change must occur to better serve the library's users.

Does this mean that librarians should all be generalists? Not necessarily. There is nothing wrong with a librarian focussing skills in a given area of public or technical service. However, that should be done in a way that keeps the service goals of the library in mind.

Benefits of This Approach

There are a number of benefits that accrue to the library, to its users and to the librarians in this new environment. They include —

- Users are better served by integration of knowledge that leads to a better OPAC and better understanding on the part of librarians.
- Improved morale on the part of librarians as a byproduct of everyone actively working together toward the same goal.
- Improved morale among technical services librarians as they are perceived to have more value and are recognized for their knowledge.
- Improved productivity in the library as everyone has a better understanding of the library mission that can serve as a focus for problem solving, decision making, etc.
- Improved public image of the library as a truly service oriented institution.

There has been a perception among librarians that public service is a more highly valued aspect of librarianship than technical service. In fact, this is a commonly presented reason why library school students shy away from technical service. There seems to be little appeal for the work in technical service. Yet, who was first responsible for automation in the library? Technical services led the way, building on the success of OCLC, RLIN, and WLN. Who gives the library its foundation in terms of acquisition of materials and organizing those materials? Clearly the technical service staff. Becoming more visible and an integral part of the library team can benefit the individuals in technical service in terms of acceptance of their expertise in areas usually foreign to their colleagues but of importance to the library. Rather than practicing their skills in the back room and in isolation from their colleagues and the public, they can become members of the service community in a way that can bring them recognition from their library peers and from the public that is the recipient of the library's service. Public service staff would benefit by a clearer understanding of the principles involved in building the OPAC upon which library users rely.

FUNCTIONS OF TECHNICAL SERVICE LIBRARIANS

Who better to answer questions of how to use the OPAC than those who built it, who made the decisions on what to include and how to include it? In the area of subject access, who knows Library of Congress subject headings and medical subject headings better than the catalogers who assign the headings? How do we use this expertise in the front line service to the user?

Technical service librarians are likely to have a number of skills that may be somewhat hidden from their public service colleagues. Technical services are often thought of as fairly narrowly defined positions — acquisitions and cataloging. But think of the various roles that are implied — evaluator, selector, acquirer, organizer, retriever, subject specialist, indexer, interpreter, equipment maintainer, workflow analyst, systems analyst, educator, trainer, manager . . . . It is time that libraries make better use of those skills and that the technical service librarians be willing to modify their traditional behind the scenes mentality and become active in sharing the service mission.

There are at least two other areas where technical service librarians have valuable knowledge — quality control and record structure. In fact, catalogers have been perceived in some quarters as caring too much about quality, at the expense of productivity. Some of that perception is legitimate. When valuable time is spent reviewing work and correcting insignificant errors while a backlog of materials purchased but not accessible exists, it is legitimate to question the priorities of technical services. However, in the area of authority control, for example, the OPAC can benefit from machine control procedures designed by catalogers who have long dealt with such issues. Quality should be the business of everyone in the library but quality of service should be the foundation upon which other quality efforts are built.

The ability to combine the cataloger's knowledge of record structure (MARC format) with the reference staff's understanding of what the public needs and how they use the library's systems to attempt to find what they need offers an opportunity to work together on user interface design. There may be opportunities to study user behavior to work toward better interfaces and improved indexing.

The truly radical approach would be to have catalogers performing such tasks as systems design and quality control, in addition to actually working with the public. Catalogers should not stop cataloging but rather they should be encouraged to spend less than a majority of their time doing that. Catalogers need to continue to catalog, but the focus should be on cataloging the most difficult materials and in managing the cataloging process and workflow. And, part of their time should be spent in communicating with those who use the catalog, gathering input on problems of use, so that cataloging decisions can be made with the user in mind.

So, who is going to be doing the cataloging when the librarians are off consulting about systems design, quality control, user problems, etc.? Paraprofessionals are an effective answer. However, the cataloging librarian has a responsibility that may not be obvious here, again relating to the library's mission. While paraprofessional staff can be very effective and efficient, what they frequently lack is an understanding of the library as a whole and as a service institution. This is especially true in technical service where they are isolated from the public and from public service. They frequently do not understand the larger environment or appreciate how their product is used. It is a part of the librarian's responsibility to teach the paraprofessional the mission of the library and to lead by example by making that mission the foundation for technical service work.

With professional catalogers responsible for training, planning, workflow, quality control, etc. — all within the overall service orientation of the library — a premium is placed on responsiveness to the public. This means a constant attempt to streamline workflow as well as to regularly question cataloging policies and priorities to make sure that the goals of the library are being served by technical services. Spending inordinate amounts of time cataloging single pieces, treating every item as a problem, obsessive checking and revision all must be monitored with an eye toward a goal of making library materials available for the library's users. Anything that inhibits that goal should be examined critically. Backlogs of uncataloged materials and cataloging decisions that inhibit access are two examples of problems that should not occur in a service oriented institution. Creative resolution of these types of problems is a library responsibility, not just a cataloging activity.

HOW TO GET THERE

Resistance to change is, to some extent, a human condition. Even those who seem to thrive on change can experience discomfort if change is not managed well. However, within the library environment, change has become a fact of life. (In fact, the almost constant state of change we face is a challenge for many in our society as evidenced by such works as Rosabeth Kanter's *The Change Masters*.)[2] Public service librarians have had to change with the advent and proliferation of new electronic forms of information. CD ROMs and electronic databases have called for rethinking of duties of the reference librarian, especially as related to the end user/ searcher. Technical service librarians too have been called upon to make significant changes as the impact of technology, shared cataloging via networks, local systems, new and different information packaging and budgets for materials but not for processing call for workflow re-examination and create administrative concern for increased productivity.

Any basic changes in the library's structure should be accomplished with proper attention to planning and communication. Not every staff member will welcome change, but the chances for success can be increased if library managers are committed to making the transition as clear and non-threatening as possible. The reasons for changes and goals to be met by any reorganization should be clearly stated. Change is necessary to meet the challenges of technology, societal change, etc. The mission of the library should be made clear — there should be a clear statement that can be discussed and understood by all library staff. Hopefully, librarians will have little real difference of opinion as to the service mission of the library other than the usual semantic debates surrounding the best possible wording of the statement.

There is rarely one clear and obvious method for achieving effective change, especially where the impact may be considerable for significant numbers of staff. Library managers must be aware of current attitudes among their staff so that problems can be anticipated and addressed before they occur. Those staff members that are more open to change may be valuable allies for the library manager as they can help to influence attitudes of other library staff.

It may be well to implement a staff development program that focuses on change. Educating the staff to the challenges of a changing environment — one that is necessitated by many outside factors — may be a basic way to begin. Any significant change is best when it is thought out and people can adjust to it in their own styles. Those who go about major changes with the attitude that people will adjust sooner or later may create unnecessary attitudinal barriers that impede the success of the effort. Investing time and perhaps some funding in a planned program will likely yield benefits in the long run.

One of the best ways of influencing staff opinions is to enlist the aid of librarians who may have worked in other libraries where structures differed. Staff should be encouraged to learn for themselves what other alternative structures exist. This can be done by talking with colleagues who have worked elsewhere, by talking with staff in other libraries and by reading library and management literature regarding organizational structure. It might be possible to have a bibliography of such readings available for staff who would like to know more about such alternatives. Staff members who have either experienced other work structures or have taken the time to read about managing change, organizational structures or other relevant topics may become a valuable resource, perhaps in leading discussion groups that explore alternatives.

With a renewed commitment to the library mission, the partnership between public and technical service librarians becomes a critical factor. There are a number of ways to accomplish a closer working relationship between technical service and public service staffs. In some libraries, catalogers and other technical services staff have spent time working in public service areas. This can vary from a regular reference schedule as is true in some smaller libraries where there is a fair amount of cross-training, to places where a technical services librarian will work in public service twice a semester. In order to become truly effective, the technical services librarian must work a fairly regular public services schedule to become comfortable in the role rather than be intimidated by it.

CONCLUSION

A service mission should be instilled in all library school students. A love of books and knowledge is not enough for budding professional librarians. Librarians also need a love of people to engender the service ethic necessary to work in the effective library. However, all too often new library school graduates who do come out with high service ideals have them blunted or frustrated by their more experienced colleagues who want the newly delivered to see the real world. And that real world is all too often full of cynicism.

A lack of respect for the library user seems all too prevalent. Instead of making the library a welcoming place, the user is often treated with condescension or as an intrusion. Why can't the user read the signs, learn how to use the catalog, find the desired item in the stacks . . . ? The focus should swing to why the user cannot use the library effectively and what we can do to help him/her to do so.

The idea is not to carve out new roles for librarians but to adapt them to make them more efficient and more flexible so that as changes occur, the library can change to meet them. There are changes—less money, more money, new types of information packaging, new technology, new and more demands on time and resources, competition for funds, competition in the information "business." If the library does not do a better job in serving its clientele, whoever that clientele is, the library may not survive the information age. Others who are wiser in ways to provide satisfactory service could take over the information environment. We have much to learn about providing service, but we have the tools to do it better. New partnerships within the library, acting with a firm commitment to the library as a service institution, can make a significant difference.

REFERENCES

1. Rockman, Ilene. "Retrospective Conversion: Reference Librarians Are Missing the Action." *Library Journal* 115 (April 15, 1990):40.

2. Kanter, Rosabeth Moss. *The Change Masters: Innovations for Productivity in the American Corporation*. New York: Simon and Schuster, 1983.

The Changing Landscape of Information Access: The Impact of Technological Advances Upon the Acquisition, Ownership, and Dissemination of Informational Resources Within the Research Library Community

Wayne R. Perryman

For at least the last fifteen years information science pundits have been predicting the rise of the electronic library. Many doomsters in the field appeared to assume that the advent of electronic access to information sources would sound the death knell for traditional print media as a viable source of current information. As an extension of this thinking, it seemed only a matter of time before libraries would become museums where books, journals, and other print formats would be venerated and preserved more for their archival or artifactual import, than for the knowledge, information, and data which they conveyed. Those in need of the most up-to-date information would simply tap into their electronic resource, without ever needing to visit their local library facility.

Clearly, what has happened during the ensuing years has not been the demise of the library as the key resource center for information, but a much more gradual technological evolution which has helped to foster the introduction of many additional avenues for

Wayne R. Perryman is Deputy Assistant Director for Technical Services and Head Librarian, Acquisitions and Serials Department, for the General Libraries of the University of Texas at Austin.

access to information both within the library and outside of it. What has been perhaps most surprising is that this evolutionary process has developed not at the expense of the book or other traditional print sources, but rather in addition to it. As a matter fact, recent figures from the Book Industry Study Group project that the book industry will grow at an annual rate of 8.3% over the next five years, with unit sales of books increasing an average of 3.1% per year during that period.[1] It should be noted, however, that this change has not come about without major, continuing challenges to the library as an institution.

This article explores the impact which this technological evolution is having upon the strategies employed by research libraries in their quest to provide access to the information and services which are needed by their constituencies. It also offers some thoughts on what library administrators, collection development professionals, acquisitions managers, and others in the field can do in order to anticipate these developments, promote those changes which would be beneficial to the user community, stay ahead of the demand for increased access to non-traditional information sources, and otherwise contend with developing technologies and information media. It is evident that if libraries do not take a proactive role in fostering, developing, and/or promulgating these changes, library users will seek access to information elsewhere.

FROM BOOK BUDGETS TO INFORMATION RESOURCES

Libraries across the country have been experiencing unprecedented fiscal demands upon their budgets which have been brought about to a large extent by the explosion of information, both in terms of the sheer volume of publishing output and the proliferation of new and varied formats. As a result of these trends, many libraries have begun to define their budgets differently. Not too long ago, it was common for a research library to have a book or materials budget, which was essentially divided between monographs and serials covering a fairly limited array of formats. Now, it is not unusual to hear librarians referring to the "information resources budget," which has been designed to acquire or provide access to information in the myriad forms which it now takes. What brought about this change of terminology and what is its significance to library administration in coming years?

Until quite recently the basic form in which information was communicated within the research library community was in either print monographic or serial publications, with a smattering of other non-book or audiovisual formats to meet specialized, limited, or recreational use needs. One of the more complicated questions which needed to be addressed with respect to the book budget at that time revolved around the appropriate ratio of serials to monographs. With the explosion of other forms of information transmission, dissemination, and access, however, many new demands have been placed on an increasingly complex budgetary scenario which has already been stressed by the spiraling costs of materials, the fluctuation in the value of the U.S. dollar in foreign markets, and the erosion of many of the traditional funding bases for universities and libraries at the national, state, and local levels. In fact, in its publication *Trends 1990*, the Book Industry Study Group indicates that it expects serials budgets in academic libraries to increase an average of 13.5% annually over the next five years, while the share of the information resources available for book purchases in those same libraries will decrease by one-third to one-quarter by 1994.[2] It is obvious that some difficult and far-reaching decisions will face research and academic libraries in the next few years.

THE DEVELOPING SCHOLARLY INFORMATION NETWORK

A pervading element driving and facilitating much of the push towards new forms of information and knowledge distribution has been the rapid development of computer technology, particularly in terms of personal computers and telecommunication networks. While this development has revolutionized the ways in which the research library community envisions its role in providing access to information, it has also revolutionized the ways in which many members of the library user community at large seek access to information. There is now much discussion of the evolving scholarly information network in a context which envisions a researcher, faculty member, or student working on a multi-tasking personal computer in his or her office, home, car, etc., while accessing a variety of informational resources, be they full-text databases, citation indexes, or a library's online catalog, as well as other automated applications such as electronic mail, telecommunications networks, computer modelling systems, statistical databases, and the like.

The advent of a fully realized scholarly information network portends at least two major fiscal implications for the research library community. First, it means that researchers, faculty, students, and other information seekers will potentially have an unprecedented degree of access to informational resources, *without ever having to set foot in a library*. This likelihood could have an impact upon the basic level of fiscal support which a library receives from a university or college administration, since the perception could develop that the library is simply no longer as important to the research mission as it has been in the past when users almost invariably had to go there in search of information.

Second, the local development of a scholarly information network may be taking place largely outside of the library's usual sphere of operation or influence, for example, in the university's computing center. In fact, with the widespread availability of online catalogs the library may simply be considered one of many nodes on that network, with little recognition of the herculean efforts which go into building and maintaining that library's database. Consequently, in order to provide access to the full range of information resources the library may be placed in the unaccustomed position of having to compete with other campus enterprises for funds which would have historically been earmarked primarily for library applications.

FORGING A NEW INFORMATION ACCESS STRATEGY

These developments have resulted in a situation where research libraries are being forced to contend with a diminution of purchasing power (in terms of flat or even lower budgets coupled with rapidly increasing costs), while concomitantly being challenged to develop or, at least, gain access to an expensive technological infrastructure. This process requires that the library make significant, ongoing capital outlays for new equipment, programming, maintenance, and support while providing access to developing formats. At the same time, the library is experiencing increased competition from other campus agencies as it vies for its share of the university's finite fiscal resources.

Given this challenging scenario, what can a library do to continue to acquire or provide access to the information and knowledge resources which are needed by the university community? The fact is that there are no simple answers to this extremely complex, yet essential, question. To begin solving this puzzle, however, libraries can strive to:

- strengthen and expand the traditional library values of cooperation and resource sharing
- offer flexible, multi-faceted services
- provide access to new and developing formats
- foster new approaches to information access
- tap into the developing telecommunications infrastructure
- pursue and develop new funding sources
- develop a common vision of the future of information technology

These ideas will be explored in more detail in the sections which follow.

STRENGTHEN AND EXPAND THE TRADITIONAL LIBRARY VALUES OF COOPERATION AND RESOURCE SHARING

During the past few decades it has become obvious that libraries can no longer afford to take the "universe of knowledge" approach to collection building. It is simply not reasonable either economically or physically to expect any one library to collect informational resources without serious consideration of available finances and local needs. Thankfully, during the same timeframe when libraries were coming to grips with economic reality, international and national bibliographic databases, such as OCLC, RLIN, UTLAS, and WLN, have developed which provide libraries with excellent access to information on other libraries holdings. This information has been used to good advantage in better coordinating collection development programs and in making more effective resource sharing arrangements.

As a prime case in point, since 1978 a number of libraries in the

Research Libraries Group (RLG) have made a concerted effort to coordinate their collection building programs in order to ensure that they were not unduly duplicating each other's holdings, particularly of expensive or low-use materials. This effort, using the RLG Conspectus, has enabled RLG institutions to identify their collecting strengths and intensity levels in a consistent and commonly understood way. While a well-conceived, up-to-date collection development policy is necessary to any library, the Conspectus approach greatly enhanced this value by extending it far beyond the confines of any given institution. Such shared information has had a major impact upon the administration of collection development activities within the RLG, as libraries have had to make serious decisions about where to make cuts in their limited budgets in order to balance the provision of essential services against the concept of ownership of information.

Another major element in resource sharing, and one which must work hand-in-hand with any cooperative collection development effort, is interlibrary loan, which is itself a venerable institution within the library community. Interlibrary loan can take many forms from the casual, occasional borrowing and/or lending which virtually all libraries have participated in, to more formal reciprocal borrowing arrangements. One pioneering example of the latter type of arrangement was initiated between the libraries of Stanford University and the University of California at Berkeley in the early 1970's. This cooperative arrangement facilitated the trans-San Francisco Bay delivery of library materials, as well as faculty members, and graduate students between these two institutions three times a day in a van dubbed "The Gutenberg Express." This arrangement has helped to assure that scholars on both sides of the bay have free and open access to these two preeminent collections, with full reciprocal borrowing privileges.

A more recent example of an innovative arrangement is a service agreement initiated in 1989 between the libraries of the University of Texas at Austin and the University of Texas at Permian Basin. This arrangement enables patrons at UT-Permian Basin to receive full-text copies of articles from the substantial holdings at UT-Austin via telefacsimile on an expedited, albeit paid, basis. Through the combined use of modern technology and forward-thinking, these two libraries can now share information required by library users on the same day despite the fact that they are separated by over 300 miles.

These are but two examples of the ways in which libraries can seek to better share information sources and thereby maximize their use of precious fiscal resources. Again, a key element in any such resource-sharing effort is a common understanding of what a particular library's collection strengths are. It is clear that, prior to entering into a formal resource-sharing relationship with another institution, all parties must be as cognizant of the needs and resources of the other as possible. To this end, a well-articulated collection development policy, whether based upon the Conspectus model or upon another format developed by a particular library, is essential.

OFFER FLEXIBLE, MULTI-FACETED SERVICES

Libraries also need to look beyond their traditional array of services for innovative ways to entice new users to the library and to retain as much of the existing user population as possible. Given the growing emphasis upon technology, this is the direction in which many libraries have looked when seeking to develop new services. For example, a number of libraries have set up personal computer centers within the library facility, offering faculty, students, and staff access to equipment which might not be readily available elsewhere on campus. In some cases, the establishment of such facilities has put the library in direct competition with other campus departments, which might be offering similar facilities, although a little competition may be healthy, particularly at institutions which serve a large, diverse, and dispersed student body population.

One pitfall of housing a microcomputer center within a library is the ongoing costs associated with maintaining and upgrading the hardware and software in order to keep the technology current. These costs put an added burden on an already taxed library budget and may not be totally manageable without the fiscal and/or logistical support of other outside sources, such as a computer manufacturer, computation center, university administration, granting agency, etc.

Another area where libraries can use automation to provide information to users is through a variety of computer-based information services. Depending upon a particular library's resources at least some of these services could be underwritten by funds from its information resources budget, while other more expensive or lesser-used offerings could be provided on a fee basis. At the University of Texas at Austin, fee-based services include full-service online searches to over 400 databases, monthly "current awareness" printouts, and end-user searching services for over 120 databases. Free services range from access to the UTCAT (The General Libraries online catalog) and OCLC databases to numerous CD-ROM workstations which provide access to a wide variety of citation indexes.

PROVIDE ACCESS TO NEW AND DEVELOPING FORMATS

One of the ongoing challenges to libraries in the 20th century has been deciding upon which formats to commit their resources to. Until recently this decision-making process was reasonably straightforward, since there were relatively few formats available, most of the more specialized formats were limited to specific disciplines, and the technology required to utilize them was relatively unsophisticated so that equipment costs were generally not a major consideration. Other than cartographic materials, slides, filmstrips, kits, realia, and ephemera, plus a proliferation of microformats, the majority of non-book formats dealt with in research libraries, both in terms of cost and quantity, were in the performing and fine arts, including various types of phonodiscs, audiotapes, and film. Print media, whether monographic or serial, was far and away the dominant means of information dissemination.

With the advent and widespread application of optical disc storage and laser technology, the digitization of audio and video production media, and the availability of advanced personal computer technology, libraries have increasingly been expected to collect many of the newer developing formats, such as compact disc, digital audio tape, laser disc, computer software, etc. All of these formats require highly sophisticated and often expensive equipment for playback or use. At the same time that they have been developing the means to take advantage of these new formats, libraries have been expected to continue acquiring and maintaining collections of older more traditional formats.

Further complicating matters has been the recent development of multi-media packages which amalgamate formats into a presentation package requiring a specially configured personal computer workstation which can be very costly to purchase and maintain. For example, the Voyager Company has initiated a *CD Companion* series of classical music which consists of compact discs and software diskettes, and requires a Macintosh personal computer with hard disk drive, a Macintosh-compatible CD-ROM player, hypercard software, with printer, headphones and/or speakers optional, in order to operate. Other offerings from Warner Communications, in a series entitled *Warner Audio Notes*, require the above equipment, plus (for use of all of the presentation features) a CD + G player (for graphics), laser disc player, videodisc interface, color TV monitor, MIDI (musical instrument digital interface), and synthesizer. Such multi-media packages are relatively inexpensive to purchase, with an offering of Beethoven's *Symphony No. 9* costing $99.00 from Voyager and a Warner series entry of Mozart's *The Magic Flute* going for $66.00 (not including the laser disc, which would cost an additional $70.00). Despite the initial low cost of these packages, the lack of availability of suitable playback equipment within the library would seriously limit their immediate usefulness to a library patron, unless they had the equipment elsewhere at their disposal.

These examples illustrate a dilemma which has become increasingly prevalent in libraries in recent years as new technology has developed at such a rapid pace. That is, determining at what point a library should begin committing its information resources to a particular format, or, as in the case just mentioned, to a multiple format compilation. The responses to the question of whether or not to commit resources to a given format tend to fall into two opposing camps, with some opining that libraries shouldn't purchase a format for which they cannot provide patron access, and others stating that the library should be on the cutting edge of technology, acquiring new and developing formats in anticipation of patron demand as soon as they have proven to be commercially and technologically viable, without immediate concern for the availability of playback equipment. The best solution lies somewhere between these two extremes — begin to acquire proven new formats as soon as possible, but only if the equipment needed to utilize those formats is expected to be available to interested patrons either in the library (if not immediately, then within the foreseeable future) or elsewhere on campus.

FOSTER NEW APPROACHES TO INFORMATION ACCESS

Another major issue which has recently come to the forefront in discussions of how best to meet library users' information needs has been the concept of viewing the provision of *access* to information as an alternative to the actual *ownership* of the information source. With their active resource-sharing activities, libraries have always sought to make informational resources not physically housed in their facilities available to their patrons. This concept has taken on a whole new dimension, however, with the increase in the variety of available computer-based information services, including citation and full-text databases, CD-ROM workstations, etc. These discussions have been further fueled by the widespread availability of direct access to the online catalogs of research institutions nation-wide, as well as by the continued growth of international bibliographic utilities. Most recently the concept of information access has been touted by some as a means through which libraries, in seeking to preserve their finite information resource budgets during an extended period of fiscal austerity, could reduce their ongoing commitments to lesser-used serial publications, while still providing on-demand access to articles contained in those publications through an electronic resource or document delivery service.

This latter concept bears further exploration since how libraries choose to contend with ever-increasing serial costs is one of the most serious questions facing the research community at large. At the same time it is in the arena of the scholarly journal where the concept of the electronic library has seemingly begun to come to fruition, with increasingly frequent allusions to the "electronic journal." Despite the more persistent discussions of the electronic journal in the literature, at conferences, and elsewhere, publishers are still unclear as to the direction which the journal market should be going. As Karen Hunter, vice president and assistant to the chairman of Elsevier Science Publishers B.V., stated recently:

> Overall . . . the number of . . . full-text journal projects [within the scientific and research publishing community] has been virtually unchanged for six or seven years. Publishers can do a lot technically, but most are uncertain what to do, and the signals from the market are *extremely* unclear [Hunter's italics] . . . Given this confusion in the market, the high investment costs, the desire on the part of authors, editors, and readers for continuity of the existing product, and the potential for product cannibalization, it is not difficult to understand the gradual rate of change in distribution technology over the past decade.[3]

Even if the pace of change has been slower than initially envisioned, there have been some recent indications that the electronic journal concept is indeed moving forward. Interestingly and somewhat ironically, one of the more successful electronic journals to come along of late is the *Newsletter on Serials Pricing Issues*, which is edited by Marcia Tuttle of the University of North Carolina on behalf of the American Library Association, Association for Library Collections and Technical Services, Publisher/Vendor Library Relations Committee, and is available on BITNET, ALANET, and the Faxon DataLinx system. Despite the fact that this newsletter has been put out under the auspices of ALA, it has gotten out ahead of existing publication policies within that organization, since, given the fact that it is now only issued online, it doesn't fit into the long established editorial oversight mechanisms which have been set up for print publications. The realization of this fact has, in turn, engendered much discussion within ALA of the issues related to electronic publishing, including ownership and copyright, editorial responsibility, distribution policies, publication costs, etc.

As the prices of serial publications, particularly in the scientific and technological disciplines, have continued to soar, libraries have begun to consider other means of providing access to the information contained in them, while at the same time attempting to identify and grapple with the causes behind the rampant price increases.

Depending upon one's perspective, spiraling serial costs have been blamed on many factors including publisher price gouging and exorbitant profits, increased publishing output (more journals, more pages per journal), increased costs of printing and materials, fluctuating foreign currency exchange rates, a lack of cohesive vision, oversight, and advocacy within the academic/library community, and the entrenched "publish-or-perish" system within academia which mandates the publication of scholarly articles and monographs in order for faculty to meet tenure requirements. The fact is that all of these factors have played a role in the increasing costs of journals, some more so than others. Publishers, vendors, librarians, and members of the scholarly community will need to continue working together to address these issues, since it is clear that the relationships of all of these parties are inextricably intertwined, and it is in the best interests of everyone to arrive at a reasonable solution which equitably balances profit motive with scholarly endeavor.

In the meantime, libraries will need to continue exploring and promoting the use of alternative access sources, particularly fulltext databases and the fledgling document delivery services, since these offer perhaps the best alternatives to actual ownership of the information sources, while exploring new concepts in information technology. One example of an experimental document delivery project which has developed within the past few years is Project ADONIS which provides access to articles contained in over 200 biomedical journals and is stored on CD-ROM; copies of articles from journals included in the ADONIS database can also be requested online through the Faxon DataLinx network.

Another ground-breaking information delivery program has been developed by the Chemical Abstracts Service which, working in conjunction with STN International (an international scientific and technical information network), provides online access to a variety of information sources including *Chemical Abstracts*, the *Registry File* (containing a listing of 9 million chemical compounds), along with 20 full-text journals from the American Chemical Society, 7 journals from the Royal Society of Chemistry, and 5 polymer journals from John Wiley & Sons.

Due to the nature of research in the scientific and technological fields, in terms of the paramount demand for current information, the use of sophisticated and/or developing technologies, and the high costs associated with traditional publishing in those fields, more experimentation with alternative information sources has taken place to date within those disciplines than within the humanities, social sciences, and fine arts. There remains a major market in potential document delivery services outside of the scientific and technical fields which is largely untapped, but which could serve to revolutionize research strategies within those fields as well. Libraries would be well-advised to promote the development of such services and be responsive to experimentation with new concepts in document delivery in all disciplines as they occur.

While there has been much recent discussion of the electronic journal, one must not neglect consideration of the electronic monograph. While journal literature might lend itself more readily to production and dissemination in an electronic mode, and indeed there might be stronger justification for having serial literature appear in that form, there are also some monographic applications which bear similar consideration, particularly where the information presented needs to be very current and/or frequently updated, or where the text would benefit from the superior indexing capabilities of an automated application.

A precedent-setting example of just such an electronic monograph is *Superbook*, which was developed at Bellcore, the Bell Communications Research Laboratory, and has been described by Bellcore project team leader and cognitive psychologist, Dennis Egan, as "an experimental hypertext document browsing tool . . . developed as a project to aid telephone service people in accessing large technical manuals, frequently updated, in electronic format, and to assist them in quickly, accurately, and completely finding relevant information in a large text."[4] *Superbook* is presently receiving on-site testing in the Albert R. Mann Library at Cornell University. This test includes an investigation of how people search out information in texts, why they don't always find what they are seeking, and how an electronic application might aid the information retrieval process. One problem which Bellcore has attempted to address in this effort is the tendency for a reader to lose his or her perspective or place in a document when viewing text in detail in an electronic or online application. To circumvent this problem, Bellcore programmers have developed what they refer to as a "fish-eye view" using dynamic tables of contents which display while the user is reading the text concerning a particular subject and indicate where that subject is represented elsewhere in the text, which in turn "allows you to see something in rich detail without losing its context."[5]

TAP INTO THE DEVELOPING TELECOMMUNICATIONS INFRASTRUCTURE

One of the driving forces behind this new access to alternative information sources has been the national telecommunications infrastructure which has developed following the development of the Department of Defense's pioneering ARPANET network in the late 1960's. The importance of telecommunications to the scholarly enterprise and, by extension, to research libraries cannot be overemphasized—it has simply revolutionized the ways in which information and data is defined, accessed, processed, and used in modern society, as well as how knowledge is created and communicated. As Ronald Larsen has put it:

> Network connectivity has now become a basic requirement for scholarly research in many disciplines, providing access to one's peers and to high performance computational resources unavailable on many campuses. The network is also becoming widely used as an access vehicle to information resources such as libraries and discipline-oriented databases. Without such a backbone in place, institutions would have had to rely upon their own resources and expertise to establish point-to-point communications mechanisms with specific other institutions.[6]

From the modest, but highly successful, beginnings with ARPANET, a highly articulated system of often overlapping national, regional, and local networks, such as BITNET, Internet, and THENet (Texas Higher Education Network), has developed which has been used to link many business, government, and education enterprises. A current development which could further speed the promulgation of the scholarly information network is a proposal before the United States Congress to set up a new high speed, high capacity network, which would be known as the National Research and Education Network (NREN). This system would have the capacity to link up many of the nation's high performance computing facilities, or "supercomputers," providing access to over 1,300 institutions within a five year period, in order to facilitate communication, research, file transfer, and advanced computing capabilities. This network would have the capacity to transmit the equivalent of 100,000 pages of printed text per second.

To remain vital to the research endeavor, libraries must continue to seek the means to insinuate themselves into this telecommunications framework, since it plays such an integral role in the evolving scholarly information network. As Kaye Gapen defined it during her keynote address at the first INFORMA conference which was held in Austin, Texas earlier this year:

> The emerging goal [of an integrated information system] is a seamless electronic environment in which individuals may access a variety of information sources in a manner that is simple and easy, and independent of time and place and subject discipline, for purposes ranging from augmenting and refreshing memory to learning, decision-making and creating or uncovering new knowledge.[7]

It is clear that such a seamless electronic environment should not and, indeed, cannot be developed without the direct involvement of the information technology specialists within the library community.

PURSUE AND DEVELOP NEW FUNDING SOURCES

As libraries continue their trek through this evolving new world of information access, pains must be taken to guarantee that changes do not come at the expense of other essential information sources or services. As previously noted, the library community is experiencing an explosion, rather than a constriction, of information resources which means that libraries will need to continue to support most of the traditional formats, while developing the means to provide access to the newer, or as yet unenvisioned, ones. To do so will clearly entail substantial additional monetary outlays.

Since the funding which has traditionally come from local, state, and national legislative sources has been increasingly difficult to garner, in light of the crush of other societal demands being placed upon those same resources, publicly supported libraries will increasingly need to explore alternative funding sources. These sources include government funded grant programs, foundation grant programs, corporate grant programs, and, of course, private philanthropy.

When one thinks of grant programs which would be applicable to libraries, the first examples which often come to mind are those funded by the federal government, such as the HEA Title IIC and IID programs or the NEH Challenge Grants Program. But, in addition to these important programs, there are literally tens of thousands of other grant makers in the United States, many of whom regularly support library programs. Sources of information on possible grant funding programs include the *Directory of Research Grants*, which is updated and published annually by Oryx Press; the *Foundation Directory*, produced by the Foundation Center, the *Foundation Grants Index*, and *Grants*, all of which are online databases available through DIALOG; *IRIS* (Illinois Researcher Information System), an online database produced by the University of Illinois; and *SPIN* (Sponsored Programs Information Network), an online database produced by the Research Foundation of SUNY.

While anyone who has worked on a grant proposal will attest to the fact that they are very time-consuming and labor-intensive undertakings, the rewards are often well worth the effort, given the monetary recompense, the prestige and recognition for the institution, and the momentum which the successful completion of a grant project can create in terms of obtaining future grant support. Regardless of these benefits, in these tight budgetary times it is virtually mandatory that any research library have a staff devoted to development projects such as this, or at least have very close ties with the university's development office.

One problem which some libraries have experienced with respect to grant funding is that grant proposals are sometimes developed elsewhere within the academic community without consideration being given to the added demands which such a grant might place on the library, e.g., necessitating the purchase of additional materials in support of a new academic program or discipline. It is important that university administrators and faculty be aware of the potential impact which a grant award they are pursuing might have on the library and build some level of library support into the proposal when necessary. By having an active liaison relationship with the university's development office or council, a library should be able to avert any such problems by ensuring that its needs or requirements are weighed and documented.

DEVELOP A COMMON VISION OF THE FUTURE OF INFORMATION TECHNOLOGY

One thread which has been woven through this paper is the need for communication. Librarians need to play an advocacy role in the many developments which are taking place in the field. This role requires involvement at all levels within the library community, the university, the political arena, and industry. As Patricia Ohl Rice has pointed out, "if universities want to embrace the electronic library of the future, they must be willing now to let librarians take an active, even aggressive role in the planning process."[8] It is incumbent upon the library community to make certain that the university administration is aware of the expertise which they have vested in their campus libraries and take advantage of that resource.

In much the same vein, librarians must continue talking to each other to develop a cohesive vision of what the future *should* hold in store for libraries. The library community is in a unique position within society and should be playing a strong hand in helping to shape the future, rather than simply reacting and responding to changes which others outside of the community are promoting to their own ends. Librarians need to send a clear signal to publishers, vendors, and other information suppliers indicating what they perceive the needs of libraries and information seekers to be and how they might best be met. In some instances this process might result in cooperative endeavors between libraries and publishers, libraries and vendors, etc.

In dealing within the political arena and attempting to restore some of the funding which has eroded from libraries and higher education in general in recent years, librarians need to dispel the long-held notions and stereotypes surrounding the library community. Politicians need to be aware of the essential role which libraries play, and will continue to play, in the academic community and the fact that they are contending with technological forces scarcely imagined 10 years ago. It is heartening to note that the concept of the National Research and Education Network (NREN) was broadened from an earlier incarnation as the National Research Network, largely in response to outcries from the education community. The fact that the focus of this project was broadened indicates that consciousness-raising efforts can have a real impact on national public policy, even when dealing within entrenched bureaucratic structures. It is most important that, when lobbying parties outside of the library and information science field, librarians work to eradicate the antiquated perception that libraries consist solely of shelves of dusty books and periodicals, and create a new 21st century view of libraries on the cutting edge of information technology and services.

In addition, librarians need to continue forming new coalitions with other interested parties outside of the profession, both within the educational community and in the private sector, to discuss the future trends and developments. The INFORMA meeting held in Austin from April 29 through May 1, 1990 is a prime example of the positive sort of dialogue which must take place if librarians are to be able to develop a clear, workable vision of the future. The purpose of this IBM-sponsored conference, as stated in its brochure, was to "provide a forum for sharing knowledge about technological advances, and explore with IBM current capabilities and potentials for important and interrelated technologies of special interest to the library community."[9] This invitation-only event brought together over three hundred library directors, high-level university administrators, and academic computing center directors from around the country to engage in essential discussion with one of the giants in the information technology field. Again, this is just the sort of dialogue which must be promoted if librarians are to play a major role in developing the future technological infrastructure which is of such vital interest to the entire library community.

Finally, librarians should not overlook the value-added services offered by a growing number of trade vendors and the need to continue communicating changing library requirements and concerns to them as well. Some of the more forward-thinking vendors, particularly those which specialize in the academic market such as Blackwell North America, The Faxon Company, and Yankee Book Peddler, can often be an excellent source of information on "the trade," assisting with cost projections, tailoring management reports to a library's needs, etc., and generally working with the library community to make everyone's jobs a little easier. Many of these vendors have also developed advanced automated systems and/or interfaces which assist libraries in obtaining bibliographic verification and price information, facilitate the flow of information on shipments of materials sent to the library, transmit invoice information, etc. As the intermediaries between the library and the publishing communities, vendors are in a unique position to help articulate the needs of all parties, and librarians should recognize and take advantage of this fact.

CONCLUSION

Society is experiencing an explosion in the variety of new information formats and resources. This change will not come at the expense of the book or most other forms of traditional print media, but rather in addition to them. If librarians take a flexible, forward-thinking view and embrace these changes, providing new approaches and services, while working to fulfill the extremely varied information needs of their user population, libraries will experienced a renewed florescence as an institution. Regardless of how many other technological resources might arise to potentially entice information seekers away from the library, librarians will remain in the special role of information mediator. As Patricia Glass Schuman succinctly stated in a recent issue of *Library Journal*:

> The mission of librarians is not just to simply fill specific information needs. Our mission is to solve information problems . . . Librarians must equate information with understanding . . . [and] distinguish between data and information, between facts and knowledge . . . Our challenge is not just to provide more information, or even just the right answers. Our challenge is to help people formulate the right questions.[10]

In order to maintain the validity and vitality of the library in the educational enterprise, librarians must play an active role in developing the strategies and mechanisms required to meet the changing information/knowledge access needs of the academic community. This involvement requires that librarians not only respond and react to changes in the information access industry, but that, as active participants, they provide clear signals, a common vision, if you will, as to which directions they think these developments should go. This effort mandates involvement and dialogue at all levels, from university administrators, faculty, students, and staff (both within a particular institution and elsewhere) to authors, publishers, trade vendors, computer hardware/software manufacturers, and other information professionals, ad infinitum. To not do so would mean the abdication of the library communities' stake in what is proving to be the most exciting, formative, provocative, and challenging period in the history of information technology.

NOTES

1. Book Industry Study Group, *Trends 1990* (New York, New York: Book Industry Study Group, 1990).

2. Ibid.

3. Hunter, Karen, "A Publisher's Perspective (Seminar on the Future of the Scholarly Journal)," *Library Acquisitions: Practice & Theory*, v. 14, 1990, pp. 5-13.

4. Sievers, Arlene Moore, "Report on the Institute on Collection Development for the Electronic Library," *Newsletter on Serials Pricing Issues*, June 18, 1990, pp. 10-21.

5. Ibid.

6. Larsen, Ronald L., "The Colibratory: The Network as a Testbed for a Distributed Electronic Library," *Academic Computing*, February 1990, pp. 22-23, 35-37.

7. Statement made by D. Kaye Gapen during her keynote address at the first INFORMA Conference held in Austin, Texas from April 29 through May 1, 1990.

8. Rice, Patricia Ohl, "From Acquisitions to Access (Seminar on the Future of the Scholarly Journal)," *Library Acquisitions: Practice & Theory*, v. 14, 1990, pp. 15-21.

9. Contained in the statement of purpose included in the information packet provided to attendees of the first INFORMA Conference held in Austin, Texas from April 29 through May 1, 1990.

10. Schuman, Patricia Glass, "Reclaiming our Technological Future," *Library Journal*, March 1, 1990, pp. 34-38.

ADDITIONAL READINGS

Baron, Joel H., *Scholarly Publishing in the 21st Century: "A New World Beckons,"* (Westwood, Massachusetts: The Faxon Company, 1990).

Devin, Robin B. and Martha Kellogg, "The Serials/Monograph Ratio in Research Libraries: Budgeting in Light of Citation Studies," *College & Research Libraries*, January 1990, pp. 46-54.

Didier, Elaine K., "A Synergistic Approach to Defining a New Information Environment," *Academic Computing*, February 1990, pp. 24-25, 38-41.

Gwinn, Nancy E. and Paul H. Mosher, "Coordinating Collection Development: The RLG Conspectus," *College & Research Libraries*, March 1983, pp. 128-140.

Hewitt, Joe A., "Altered States: Evolution or Revolution in Journal-Based Communications?" *American Libraries*, June 1989, pp. 497, 499-500.

Hewitt, Joe A., "Objectives of the Seminar (Seminar on the Future of the Scholarly Journal)," *Library Acquisitions: Practice & Theory*, v. 14, 1990, pp. 1-4.

Kaser, Richard T., "The Future of Scientific Communication — The View from Chemical Abstracts Service (Seminar on the Future of the Scholarly Journal)," *Library Acquisitions: Practice & Theory*, v. 14, 1990, pp. 31-42.

Molholt, Pat, "Libraries and Campus Information: Redrawing the Boundaries," *Academic Computing*, February 1990, pp. 20-21, 42-43.

U.S. Congress, Office of Technology Assessment, *High Performance Computing and Networking for Science — a Background Paper*, OTA-BP-CIT-59 (Washington, D.C.: U.S. Government Printing Office, September 1989).

Issues in Bibliographical Access and Control in the Online Information Environment

Olivia M. A. Madison

INTRODUCTION

The online library information environment is changing rapidly, with events potentially out-distancing the capabilities of libraries to respond, financially and technologically. Within this changing environment, greatly expanded access and effective delivery systems will become the essential components of successful library-provided online systems. As online systems become more sophisticated, the relationships between these two components will become increasingly transparent to the users of library systems. If libraries fail to respond to the growing informational demands associated with access and delivery, their roles within the information environment will slip through their grasps. This indeed would be a tragedy, as I believe library users look to the resources of libraries as primary information sources.

Online library automation in the 1970s and 1980s primarily involved automating technical services functions. The first functional tool automated tended to be the local public access catalog, usually consisting of a "converted" or partially "converted" card catalog, containing monographs, and often coupled with a machine-readable serial title database (usually previously disseminated for public access in a book or a microform format). Initially, this online tool has tended to be a stand-alone resource and, when integrated with other online systems, it has been done so with circulation and/or acquisitions systems. The 1990s will increasingly see a library's locally-

Olivia M. A. Madison is Head of the Cataloging Department at Iowa State University Library, Ames, IA.

95

created and maintained database of bibliographic records as one component of a complex array of information databases and document delivery systems. These will all be connected through complex layers of electronic networks. Librarians have clear responsibilities in meeting the challenges associated with local online systems and the information networks connecting those local systems with the outside world of electronic information.

The following article will explore the challenges of bibliographic access and control associated with expanded public access to local library collections. These challenges relate to the increased need for descriptive analysis; the growing interrelationships between bibliographic description and access; and the implications of the growing confederation of externally-accessed online public catalogs in relation to networking. However, before I begin, I will describe briefly the rapidly changing local and external computing environments facing library automation to put into context my comments regarding important issues in bibliographic access and control in the future online information environment. The following description clearly is a summary and is not meant to be a complete historical review of recent past events.

THE FUTURE OF ONLINE LIBRARY SYSTEMS

While what I refer to as the future of computer networking in some capacities has already arrived in a few libraries, the computing and networking technologies required are still beyond the reach of many libraries. The access and delivery issues raised through the current and future use of these systems are of interest to all libraries, since how these issues are dealt with now will ultimately affect the future of those libraries that will follow.

As mentioned before, local online public access catalogs (OPACs) are just one type of online database that library automated systems may support. However, I believe these catalogs should serve as the foundation for the design of accessing other databases. I believe this for two reasons. First, the online catalog provides access to actual library holdings, holdings that library users have immediate access to and therefore have the most interest in. Second, the MARC communication format's highly sophisticated design has proven successful in supporting a variety of different types of information. The format can be manipulated in complex ways, thus providing means of access that should serve as the basis for how users want to access other databases available in automated systems and networks.

If they are to meet the demands of tomorrow, future online systems should have the capabilities of serving as a front door so to speak for a myriad of databases, both locally mounted and externally accessed. One of the most challenging decisions facing librarians today is that of selecting those databases and designing the networking architecture that will serve user needs. The ramifications of these choices have critical consequences to available resources, whether they be dwindling library operating and acquisitions budgets, increasing demands on library staffing, or expensive computing operating facilities.

Providing access to locally-mounted databases has great interest to those libraries with powerful computing system capabilities. There is a great variety of databases that libraries may choose to mount on their local systems, whether locally or commercially produced. Obviously libraries usually first mount their local OPACs, with their component systems of circulation, acquisitions, and serials control. Libraries are now purchasing subsets of the commercially-produced analytical databases on CD-ROM with stand-alone access for literature searching. While these CD-ROM-based systems are increasingly widespread in application, their major drawback is their sometimes infrequent updating with current analytical data. Some libraries are now promoting the use of the CD-ROM databases in conjunction with online end-user searching through systems such as DIALOG, BRS and Wilson.[1] Except for those few libraries that have already done so, the next step for libraries is to mount both commercially-produced and locally-produced analytical databases onto their local systems. While examples of the commercial databases are well known (e.g., ERIC, AGRICOLA, Humanities Index, MEDLINE, etc.), the possibility of also loading locally-created indexes (e.g., indexes for local newspapers, local manuscript indexes and abstracts, etc.) is equally exciting. The next logical step for accessing these information tools will be to interface these "mounted" analytical databases with OPAC holdings. The Colorado Alliance of Research Libraries (CARL) System has made innovative progress in developing a system that indexes the table of contents of serials actually held by its member libraries and loads the resulting databases onto local OPACs.[2] A second type of database to mount locally is externally-created MARC files, such as GPO records, Center for Research Libraries records, or Landmarks of Science records. These databases could either be accessed separately from the local cataloging database or added directly into the local cataloging database. A third type of database to mount locally consists of non-bibliographic data. These databases would not usually be integrated within the local cataloging database, but rather be part of one's "menu" or choice when first accessing the OPAC. Examples of such databases are historical tax records, full text technical reports or abstracts of articles, graphic and musical materials, encyclopedias, statistical census tapes, etc.

In addition to locally-mounted computer files, automated systems have the capability of serving as gateways to externally-accessed computer files such as those described above. To access these remote computers with their computer files, sophisticated computer networks have become a necessity. From the late 1960s to the present, the federal government has been integral to the development of a variety of inter-related computer networks. The impetus for the federal government, primarily through leadership of and funding by the Department of Defense and the National Science Foundation, was the need to develop a computer network for the government's computer telecommunication needs. The resulting network was created in the 1970s and was named ARPANET. ARPANET finally split into two networks, a military network and a civilian network, the civilian network being called NSFNET. NSFNET along with the inter-related regional networks is named INTERNET. INTERNET's financial future was secured in the 1980s when the National Science Foundation decided to expand and enhance INTERNET in order to network five supercomputer centers.[3] The infrastructure of INTERNET has been tied primarily into the higher education community and is the dominant network used for its inter-institutional communications.[4] Currently, according to Vinton Cerf (the "inventor" of the INTERNET protocols) in the paper he gave at the ALA LITA's membership meeting at ALA's 1990 Midwinter Meeting, there are over 200,000 computers in 35 countries connected through INTERNET. University library OPACs are an integral component of INTERNET, with 70 OPACS now being accessible. In addition, the President of OCLC, Inc., K.

Wayne Smith, recently announced that the National Science Foundation has agreed to allow OCLC to make EPIC, the OCLC reference database, available through INTERNET.[5] The availability of EPIC through the INTERNET is a valuable resource tool that will provide broad international coverage of library resources.

The limitations of the INTERNET, technological and its membership criteria, have brought to national attention the critical need for a national high-speed computer network that would benefit a broad cross section of the American population. Legislation, originally introduced by Senator Albert Gore (Chairman of the Commerce, Science and Transportation Subcommittee on Science, Technology and Space), is now before the U.S. Congress that will support a proposed National Research Education Network (NREN). The proposed NREN is a high-speed electronic telecommunications infrastructure that would provide universal access to "high performance computing tools, data banks, supercomputers, libraries, specialized research facilities and educational technology."[6] A number of library organizations, such as the American Library Association, the Association of Research Libraries, The Library and Information Technology Association, and the U.S. Commission on Libraries and Information Science, worked hard to build into the legislation the stated need for library resource and information services.

The NREN is often referred to as the information superhighway of tomorrow. The crucially-needed NREN legislation requests $400,000,000 to fund NREN research, development and backbone facilities.[7] To support its creation and funding, the American Library Association Council endorsed an NREN resolution, which originated with the Library and Information Technology Association. This resolution not only endorsed the concept of the NREN but it also resolved "to work to improve legislative and other proposals to increase opportunities for all types of library participation and leadership in, and contributions to, the National Research Network."[8]

ISSUES OF BIBLIOGRAPHIC ACCESS AND CONTROL

The resounding implication of all these exciting developments, which automation is bringing to our technological doorstep, is quite simply the clear and overwhelming need for a dramatic increase in

timely and complete access to information, whether obtainable locally or not. Important issues regarding this increased access have
been raised with the implementation of local online systems and
their future promise of expanded capabilities to network beyond the
confines of those local online systems. Some of these issues predate
automation, in fact date back centuries, and other issues have
evolved because of the access and record manipulation that our
OPACs can provide. In particular, for the purposes of this article,
librarians now need to question and explore what should be the
analytical depth of local OPACs, with their bibliographical records
and alternative access tools. Also at issue is the diminishing boundary between description and access. The breaking down of this
boundary has occurred primarily due to the increasingly sophisticated keyword indexes and boolean searching techniques available
in many OPACs. Finally, what are the issues related to the changing networking environment? Are we redefining where we are going, as a nation, with the expanding network of bibliographic information and holdings? And, because of the nature of national access,
is the need for standardizing the designs of library OPACs of growing importance?

A. Analysis

As libraries implement online public access catalogs, librarians
face users continuing historical questioning of what public access
catalogs contain. Commonly-asked questions concern what the circulation status is of a particular title, whether or not the database
contains journal articles, what the acquisition status is of a book
recently ordered, etc. Expectations of what OPACs provide are increasing, expectations that libraries are often unable to meet, but
which in the future they might. Up to this point OPAC enhancements usually contain the automation and integration of library
processing functions such as circulation, serials control and acquisitions. Of increasing importance is increased access to library materials, in particular analytic access. The issue of analytic access to
bibliographic records is not new, but recently several authors and
speakers have expressed grave concern over the usual minimal level
of bibliographic and subject access to library collections provided
for public access catalogs. With increasing regularity the call is

raised within the library profession for an increase of bibliographic analysis as a partial solution to what is seen as minimal-level access to our monographic and serial collections.[9,10,11]

Current estimates for academic research libraries are that their public access catalogs, whether they be online catalogs, card catalogs, book catalogs, etc., provide access to only 2-3% of their library materials.[12,13] Bibliographic control and access largely accounts for only monograph and serial "containers," or what Doralyn Hickey calls "description of physical objects—books, a run of periodical volumes, a record album, a film . . . "[14,15] The individual "works" contained within these physical objects are not routinely described or given analytical access. These "lost" individual works may be individual articles within serial issues, individual chapters in books, individual issue or volume titles of serials or monographic sets, etc.

Analysis has a very old tradition in western librarianship, having many strong proponents through the centuries. The 1605 Bodleian catalog, "often called the first printed general catalog of a public library" contained a large number of analytical entries.[16] The numerous analytical entries to large sets were apparently a concession by Thomas Bodley to the wishes of Thomas James, the Bodleian librarian who prepared the catalog.[17] Analysis continued as part of the cataloging and indexing process through the eighteenth century, though was not necessarily commonplace.[18] During the late nineteenth century Charles Cutter stated that analytics are vital assets, without which many expensive works would "stand upon the shelves untouched."[19] Furthermore he did not understand how any person familiar with "the difficulties of research" could oppose the preparation of analytics, and "in fact" did not know anyone "so benighted."[20] Today, when libraries do analyze serials and monographic sets, they usually follow Charles C. Jewett's criteria found in the 1843 catalog at Brown University—"for whatever fills a whole vol."[21]

1. Externally-Created Analysis

As analytical cataloging decreased over time in the United States, the commercial sector took over the indexing and abstracting of commonly-held serials. The resulting commercial analytical databases have evolved into vital supplementary access tools to our collections. While historically these databases have been published in printed formats, analytical bibliographic vendors, such as DIALOG, BRS, and ORBIT, now provide online access to them. In addition to their published forms, libraries are now purchasing them in machine-readable form, through CD-ROM subscriptions or through the purchase of the subscription tapes. Analytical databases tend to be subject oriented, and problems have evolved over each having "different search commands and continuously changing structures. Contributing to this confusion are the almost daily additions of new sources by database wholesalers."[22] Perhaps the most serious drawback to the CD-ROM and subscription tapes for loading into local systems is that the files usually do not represent long retrospective runs—a five year historical file is not unusual. In addition, the machine-readable versions still require a two-step search. First one searches the tool, and then one searches the library public access catalog to ascertain whether or not the local library in fact "holds" the title in question. Advances in the computing technology of OPACs should enable libraries to link locally-mounted online analytical databases to the bibliographic records and their holdings information contained in the OPACs.

2. Local and Cooperative Analysis

In addition to the desired linkage of commercially-produced analytical databases, there are ways that libraries, individually and cooperatively, can and should enhance the bibliographic access to their collections. This proposed increased analysis poses questions regarding the relationships between access and description. This particular issue was recently discussed in the debate over the MARC tagging of series, with the question being whether a "descriptive" MARC tag should also serve as an "access" tag. While the decision was to retain the MARC tag enabling a cataloger to use the 440 field to both describe a series and provide its access, the questions that it raised will continue. The crux of the debate is the relationship between access and description in machine-readable records. To accomplish economically a substantial increase of access to library collections, collectively the cataloging community will need to employ a mix of different options, options that currently exist in current cataloging practice and options that will evolve out of automated technologies.

ANALYTICAL OPTIONS SUPPORTED
BY CURRENT CATALOGING PRACTICE

The current edition of the *Anglo-American Cataloging Rules* provides for a number of options for the analysis of our monographic and serial titles, and the various MARC formats support the majority of those options.[23] Many of the mechanisms are being used in current cataloging, and only an expansion of such practices would contribute greatly to the analytical depth of our cataloging databases.

The simplest analytical approach is to provide a note listing the bibliographical contents of the work to the cataloging record for the work being cataloged. The listing ordinarily would give just the "titles" or the "authors and titles" for the individual parts, whether they be book chapters, individual musical works, individual volumes of monographic series or serials with titles not dependent on the title of the "comprehensive" work, etc. Obviously, to achieve consistency in the forms of headings, it is preferred that analytical added entries be included as well as the descriptive notes. Unfortunately, for many bibliographic records, these two levels of descriptive and access are not done routinely, and such access is left to the commercial disciplinary indexing agencies.

The most expensive method to analyze any particular work is through the creation of a full analytic cataloging record for its component parts. These individual component parts can be issued physically as individual pieces or contained within a physical piece with other parts. Examples are numerous: individual conference proceedings or professional directories contained in serial issues, individual novels contained in anthologies, individual monographic titles published as part of monographic series that are classified and cataloged under the series titles, archival subcollections, etc. For these types of works, the "note" approach will not suffice. Clearly, full analytic cataloging is a very expensive proposition, especially because it so often represents original cataloging with a difficult level of subject analysis. The difficulty lies in that the subject topics are often new and therefore do not correspond to any current Library of Congress subject headings. In addition, this type of cataloging also requires authority work and control for the name headings established. Clearly libraries are very selective over the amount of analytical cataloging records they create. For those libraries that do routinely create analytic cataloging records, there exist a variety of criteria for when one does or does not analyze. One common criteria is that a library only analyzes a serial or monographic set if each individual physical piece comprises the total analytic work. As mentioned above, this was a criteria favored by Jewett. Another common criteria is that a library only analyzes a multi-part item (e.g., serial, or monographic set) if each physical piece is "analyzable," or in other words, each piece actually has an individual title associated with the piece. When libraries do not follow these commonly-applied practices, analysis involves expensive processing. In 1985, due to a growing backlog of analytic cataloging requests and the need to understand the staffing implications involved with doing the cataloging, Iowa State University Library's Cataloging Department staff sampled its backlog of analytic titles waiting to be cataloged. For those items that comprised an entire volume of a "fully analyzed" serial or monographic set (i.e., each issue or volume is analyzed), cataloging copy was found in the OCLC online database for about 80% of these titles (of which only 10% represented Library of Congress cataloging). For titles published in what was considered "partially" analyzed serial or monographic set titles (i.e., not each issue or volume is analyzed), the OCLC hit rate plummeted to 20%. In order to provide some access to this material, the Cataloging Department began to provide minimal-level cataloging for those analytic titles contained in partially-analyzed serials and monographic sets, which, after six months, had no corresponding cataloging copy in the OCLC Online Union Catalog. The Department was then able to keep current with its new receipts of analytic requests, with at least some level of access being provided.

Individual libraries can only hope to meet the growing expectations for enhanced analytical access through a major national cooperative effort. Added to just the staff expenses related to the description and access part of the cataloging process, this type of cataloging often requires subject expertise to do the subject analysis. Just as the OCLC Online Union Catalog was built to provide cooperative cataloging to "regular" titles, so should it and the other bibliographic utilities be enhanced to contain a greater proportion of analytic cataloging for those titles already represented by "container" records.

ALTERNATIVE APPROACHES
FOR ANALYTICAL ACCESS

Clearly with the potential capabilities of online systems to not only manipulate but also to store data, libraries have opportunities to rethink local cataloging practices. However, within this "rethinking," practically the context will be bound irretrievably to the MARC format through the millions and millions of existing MARC records. Just in terms of OCLC records, there are over 355 million location symbols attached to twenty-one million bibliographic records contained in the OCLC Online Union Catalog.[24] Then ponder all the individual records created through the RLIN, WLN, and hundreds of regional systems. There can be little thought of going completely "back to the drawing board." Rather, we need to be looking at enhancements to the MARC format records we have and creating ways of interfacing those records to complementary analytical information.

At the national level, OCLC, Inc. is considering a major project of enhancing its database by having table of contents information added to monographic bibliographic records. The proposal was recommended by the OCLC Users Council, with the obvious intent of improving access to monographic records. While individually, the task of adding contents information and analytical access points to our cataloging records seems daunting, collectively it would be manageable.

John Duke has proposed a innovative tripartite structure of cataloging records, which distinguishes description from access.[25] He proposes creating a potentially three-tiered structure for any given cataloging record. The first tier, a document surrogate, would contain information similar to that currently contained in our bibliographic records. The second tier of the record would be a document guide and contain a description of the content of the work, such as content information, indexes, and summary or abstracting notes. The purpose of this information would not only be to provide descriptive information but also to provide access. The third tier would be the actual text or parts of the work. According to Duke, the first tier would contain controlled headings, thus meeting the collocation objective of our catalogs. The second and third tiers would include "uncontrolled" access, in large part due to economic constraints associated with authority control. This proposal could be implemented in part with a machine restructuring of the current MARC records. The type of information contained in the second tier is already "tagged" by unique MARC variable fields.

A simpler or perhaps transitionary implementation of Duke's proposal is to index these "non-controlled" descriptive fields only in keyword indexes as opposed to our "regular" indexes. The purpose of this access delineation is to structure clearly the role of our authority systems within our OPACs. Thus creating structure to our controlled access fields as opposed to our access not regulated by our authority systems.

Greater analysis will not only provide superior access to a library's resources, it will also provide greater efficiencies. As Herbert H. Hoffman and Jeruel L. Magner noted, it will help avoid unnecessary purchases due to increased subject access and, avoid unwanted duplication for those works contained within comprehensive "titles."[26] A national cooperative effort will be a necessity if libraries will be able to meet the challenges of increased analytical access to their collections.

B. Impact of Increased Access

The sheer wonder of technology's impact on automated library systems is that we can begin to look at the content and structures of bibliographic records in ways not previously imagined. In doing so, it is necessary to continue focus on a number of issues related to what technology should provide, not what it can, and how to provide access to library resources, resources that are now not necessarily held locally.

I believe that the purposes of OPACs are not different from the purposes of the card catalogs. While one major purpose of the catalog is to identify library holdings, another is to collocate. Collocation must remain an objective in our automated systems. For example, one should be able to find all works written by a specific author under one form of name for that author. The cataloging community has been steadily moving to standardization in all aspects, whether through descriptive rules, forms of headings or subject analysis. In large part, standardization was brought about because of the desperate need for cooperation among libraries to maintain processing costs. Now that remote access to public access catalogs is generally available, that standardization is increasing in importance.

With the implementation of OPACs, the decisions as to what component parts in cataloging records are indexed have begun to cause problems with the concept of collocation. Fields that were designed to be descriptive, not authoritative, are now being indexed in order to provide greater access — content notes and summary notes are common examples. The first question asked when indexing these fields is how do you index them? For contents notes including only title information, treating these as keyword title searches makes admirable sense. However, when they also include author citations, they cause problems if the notes are meant to be title access points, or even if the fields are searched generally as authors, the form found in the contents note may not correspond to the form used as authenticated name access points. The collocation problems associated with combining the searching of names and subjects in controlled and non-controlled fields are legion. Related concern involves the on-going goal of creating systems of access that provide manageable and understandable index files. In the card catalog it was the issue of twenty to thirty card drawers containing records with the same heading. The OPAC equivalent is completing a search with 2,000 matching "hits." Searching non-controlled fields will magnify the problem which already exists in large computer databases. As bibliographic description expands from the purpose of identifying known items to also indexing, in a sense, our controlled heading structures, this concern grows geometrically.

I am not suggesting that non-controlled fields not be indexed, however librarians should be clearly aware of the pitfalls associated with them, and try to minimize them when possible. In particular, increased levels of authority control, greater specificity in controlled access points (for description and subject analysis) and more sophisticated OPAC displays indicating the structure of retrieved records would help with the minimization.

C. The "National Network"

More as an observation than anything else, I believe that the rapidly changing networking environment has changed the library community's concept of "the national network." Indeed, the concept of the national network is evolving or should I say "mutating?" It is clear that we are progressing along two paths at the moment, two paths that are meeting distinct and different needs.

Until the advent of bibliographic utilities, such as OCLC and RLIN, our ventures in bibliographic cooperation were confined largely to sharing our local cataloging records with the Library of Congress through the *National Union Catalog*. While OCLC initially was used as a cooperative cataloging database, its uses for national and international resource sharing through its dynamic interlibrary loan (ILL) system and serial union list function have grown geometrically. After only ten years, the twenty-six millionth OCLC ILL request was logged by Burling Library of Grinnell.[27] The ILL Subsystem averages 82,000 ILL requests each week, with more than 4.28 million ILL requests processed in 1989.[28] The OCLC Union List is used by over 100 groups, which represents more than 7,000 libraries, and stores the five million holdings records used to create and maintain union lists of serials.[29] The strength of OCLC is the dramatic wealth of its database; as of April 1990 the database contained nearly 21 million bibliographic rec-ords.[30] Increasingly, large non-OCLC academic research libraries are loading their bibliographic records into the OCLC Online Union Catalog as well as participating in their own individual utilities. While regional bibliographic databases are continuing to be created, for all practical purposes, the OCLC Online Union Catalog has become the de facto national union catalog, with its own semblance of standardization, that being Library of Congress cataloging practice. In addition, the wealth of its database extends beyond the U.S. borders, with OCLC's contractual agreements to load foreign national libraries' records and its growing number of foreign OCLC member libraries.

At the same time the library community is building this de facto union catalog, primarily for resource sharing through the OCLC Interlibrary Loan Subsystem and the OCLC Union List, and shared cooperative cataloging, a loose confederation of accessible online bibliographic databases is being built. While a small number of these OPACs are available through INTERNET, the number is growing. And, if the intent of the NREN is carried out, the number of accessible OPACs will grow dramatically. The advantages of accessing these OPACs are clear. First, unlike the capabilities of the OCLC Online Union Catalog, in many cases holdings information is available, information invaluable for resource sharing. General availability of this holdings information will increase the accuracy of ILL requests in conjunction with initial searching on OCLC. In addition, the title holdings information may be more complete in a library's local OPAC than through a utility such as OCLC, due to varying methods used for retrospective conversion projects and minimal level cataloging. Also local OPACs could have an authority control structure that is more complete than the Library of Congress's authority file for local and state names and more appropriate for the local file. For library users who do not have access to OCLC, they could have immediate access to remote library collections and can be quite independent in deciding the usefulness of other libraries' collections. Mary Engle, in a paper she gave at the Annual 1990 LITA Online Catalogs Interest Group meeting, described the rapidly growing number of searches taking place on MELVYL (University of California Library Systems, online union catalog) through its capability of searching the thirteen remote OPACs accessible through the MELVYL system.[31] If the experience of the University of California is transferable, it is certain that as access to this type of computer telecommunication networking is increased so will the interests of library users in this type of remote access.

The access to remote online OPACs is certainly not without its drawbacks, or should I say frustrations. Many problems are due to the lack of standardization in general OPAC design. Certainly searching command structures differ in the various systems, and even when one is searching another library's OPAC which is by the same "manufacturer," the choices for what is indexed can differ greatly, thus causing bewilderment and confusion. Help screens are not always of great assistance, and the searcher probably has no assisting documentation. Also, at present, data communications are uneven at best in terms of error-free transmissions and speed.

Before there can be overall positive satisfaction with searching remote OPACs through general communications networks, continued serious discussion and agreement on design standards for command language and generally-accepted indexing structures is needed. The Technical Standards for Library Automation Committee (TESLA) of the Library and Information Technology Association is currently working in this area. Following the examination of results of a recent survey sent out to numerous OPAC vendors and developers, the TESLA will decide on "the desirability of developing standards or guidelines in the area of OPAC indexing."[32] When OPACs use expanded or alternative command language and indexing structures, then perhaps this could be clearly stated in help screens designed specifically for non-local remote users. Ultimately problems such as these will be more transparent as we install more sophisticated interfaces. Finally, if and when the NREN is established with its stated goals, we should have available high speed, error-free data transmissions.

CONCLUSION

The primary goal of library OPACs should be the ability of users to retrieve, with minimal effort, records for accessible materials that meet their information needs. To achieve this goal libraries must expand the analytical access (with the resulting subject access) to their collections. As mentioned above, this will require an expensive mix of different methods of access, externally as well as locally and cooperatively driven. The local implications extend to financial and staff resources. The costs of providing access to commercially-produced analytical databases are high, but the resulting access is a necessity. An increasing part of libraries' financial and computer resources will have to be diverted to computer access to these databases. Also, cataloging staffs must cooperatively provide the additional access to local, regional, national and international bibliographic databases. Libraries spend major parts of their operating budgets on acquiring materials; they must also spend dollars to allow their users to locate and use those materials. Ultimately, as document delivery services become more feasible, enhanced access will grow in importance.

In addition, librarians must carefully study how machine-readable records are being used, and maximize their usefulness through a thorough understanding of their content and design. Above all, providing greatly expanded access must not be allowed to create useless access due to the magnitude of potential index retrieval. Coupled with the need for increased access is the need for greater specificity of access points. Finally, it is only through the study and understanding of the access systems now being created and used, in terms of the machine-readable records themselves, the organizational structure of those records and related authority records, and system capabilities, will the library profession be able to enhance their usefulness and place itself permanently in the information age of the 1990s. Finally, the future as expressed in this article is brilliant in its endless possibilities, and dim in the capabilities for all libraries to bask in that brilliance.

NOTES

1. Susan K. Charles and Katharine E. Clark, "Enhancing CD-ROM Searches With Online Updates: An Examination of End-User Needs, Strategies, and Problems," *College & Research Libraries* 51, no.4 (July 1990):321.

2. Gary M. Pitkin, "Access to Articles Through the Online Catalog," *American Libraries* 19, no.9 (October 1988):769.

3. Edwin Brownrigg, "Developing the Information Superhighway: Issues for Libraries," p.3. Issued as part of: *LITA Information Packet on the Proposed National Research and Education Network (NREN)* (Chicago: American Library Association, ©1990).

4. Clifford A. Lynch, "Linking Library Automation Systems in the INTERNET: Functional Requirements, Planning and Policy Issues," *Library Hi Tech* 7, no.4 (1989):8.

5. Connee Chandler, "Oldest Living Networker(?) Addresses Membership Meeting," *Action for Libraries* 16, no.7 (July 1990):2.

6. *NREN: The National Research and Education Network* (Washington, D.C.: Coalition for the National Research and Education Network, ©1989), 9.

7. Ibid., 13.

8. Carol Henderson, "Federal Development of a National Research and Education Network: A Chronology of Significant Events and Library Community Involvement." Issued as part of: *LITA Information Packet on the Proposed National Research and Education Network (NREN)* (Chicago: American Library Association, ©1990).

9. Sheila S. Intner, "Functional Inaccessibility in Libraries," *Technicalities* 9, no.12 (December 1989):5.

10. Mark Kibbey, "Carnegie Mellon University" (paper presented at the LITA Online Catalogers Interest Group program "The Evolving Catalog." 1990 Annual American Library Association Conference, Chicago, June 24, 1990).

11. Herbert H. Hoffman and Jeruel L. Magner, "Future Outlook: Better Retrieval Through Analytic Catalogs," *Journal of Academic Librarianship* 11, no.2 (July 1985):151-153.

12. David Tyckoson, "The 98% Solution: The Failure of the Catalog and the Role of Electronic Databases," *Technicalities* 9, no.2 (Feb. 1989):10.

13. Kibbey, "Carnegie Mellon University."

14. Hoffman and Magner, "Future Outlook," 153.

15. Doralyn J. Hickey, "Theory of Bibliographic Control in Libraries." In: *Prospects for Change in Bibliographic Control: Proceedings of the Thirty-eighth Annual Conference of the Graduate Library School, November 8-9, 1976.* (Chicago: The University of Chicago Press, 1977):34.

16. Sidney L. Jackson, *Libraries and Librarianship in the West: A Brief History.* (New York: McGraw-Hill, 1976):160.

17. Ibid., 16.

18. Ibid., 264.

19. Ibid., 384.

20. Ibid.

21. Ibid., 382.

22. Richard Joseph Hyman, *Information Access: Capabilities and Limitations of Printed and Computerized Sources.* (Chicago: American Library Association, 1989):137.

23. *Anglo-American Cataloguing Rules*, 2nd ed., 1988 Revision. (Chicago: American Library Association, 1988):299-302.

24. *OCLC Catalog of Products and Services* (Summer/Fall 1990):8.

25. John K. Duke, "Access and Automation: The Catalog Record in the Age of Automation." In: *Conceptual Foundations of Descriptive Cataloging.* (San Diego: Academic Press, 1989).

26. Hoffman and Magner, "Future Outlook," 153.

27. "Grinnell College, Burling Library Logs 26 Millionth ILL," *OCLC Newsletter* 185 (May/June 1990):6.

28. *OCLC Catalog*, 8.

29. "Resource Sharing," *What's New at OCLC* (Summer 1990).

30. *OCLC Catalog*, 8.

31. Mary Engle, "Electronic Paths to Resource Sharing" (Paper presented at the LITA Online Interest Group Meeting program "The Evolving Catalog: Expanding the Breadth and Depth of the OPAC." 1990 Annual American Library Association Conference, Chicago, June 25, 1990).

32. Katharina Klemperer, "TESLA on OPACS," *Library Journal* 115, no.13 (August 1990):10.

Technical Services in Public Libraries

Kenneth John Bierman

INTRODUCTION

The words "public library" bring differing images to mind to different people depending on their backgrounds and experiences. For some, the image is the small town public library that serves relatively few people with very limited budgets (sometimes as little as a few hundred dollars). Thousands of these libraries dot the American landscape. For others the image is the very large megatropolis public library that serves millions of people via a well developed system of branch or neighborhood libraries and a large, central research library. A handful of these kinds of public libraries exist (Boston, Chicago, Los Angeles, etc.) and they are, in reality, two libraries in one — multibranch library systems with all the problems associated with that type of library and large research libraries with all the problems associated with that type of library. For most people, the image of the public library is the medium-sized library that serves more than 100,000 people, has a budget in excess of a million dollars, and has multiple branches. Hundreds of these libraries (or library systems) exist and together they serve over half of the population of the United States. These libraries have a central, main or downtown library that provides centralized administrative and technical services support but, while they have larger and more in-depth collections and service than the branch libraries, they do not have, and never will have, comprehensive research collections. The branch libraries provide popular and heavily used collections and

---

Kenneth John Bierman is Assistant University Librarian for Technical and Automated Services at Oklahoma State University Library, Stillwater, OK.

services. This is the public library image that I have as I write this article.

Public libraries come in all sizes and descriptions but they all have many things in common. All public libraries, to varying degrees and emphases, provide educational, recreational, cultural and informational services including multi-media collections and programs. They all serve all ages (literally from cradle to grave) and all sorts and conditions of humanity from wealthy business persons and housepersons to the homeless castoffs of society. They serve people with varying educational, emotional and intellectual levels and with a variety of language and cultural backgrounds. In short, the clientele of the public library is the entire spectrum of society.

THE PRESENT

What can be said about public libraries in 1990 in terms of technical services and technology? In what ways are technical services provided in a public library different from technical services provided in an academic library? The following generalizations apply to varying degrees to most public libraries, although they are written with the medium-sized multi-branch public library in mind. Like all generalizations, they do not apply completely to any library but are rather summaries of reality.

Public libraries have centralized technical services divisions that typically consist of an acquisitions (often called order) department and a cataloging department. Physical processing may have its own department or the work may be performed in either or both of the other departments. Special formats (serials, non-print, documents, etc.) usually do not warrant separate department status because they are not sufficiently abundant although they may have specialized staff assigned to them within the acquisitions and cataloging (and processing) departments. Foreign language materials are not nearly as troublesome as in a research library because fewer are purchased and the only languages purchased in any quantity are languages prevalent in the community for which educated paraprofessional staff is available. In comparison to academic libraries, relatively fewer professional librarians and relatively more paraprofessional staff are employed in technical services in public libraries.

In addition to the traditional functions of acquiring, cataloging and processing library materials, the technical services division of a public library typically includes automation management. Very few public libraries have assistant directors for automation, systems, technology, etc. as is fairly common in research libraries. In addition, the delivery or courier service is sometimes included within the technical services domain in public libraries and this function may also include responsibility for system-wide supplies, etc. Collection development (i.e., selection) may be included within technical services in public libraries but more typically it is not.

Much emphasis is placed on physical processing not for long term preservation but rather to enhance the looks of the item on the shelf and to increase the number of possible circulations before the item falls apart. Covering dust jackets with plastic covers is considered an essential activity largely because the books look so much nicer (i.e., increased marketability). This emphasis on physical processing (including mending of heavily used material) has budget and staffing implications for technical services in public libraries.

Although public libraries have a great variety of media (audio cassettes, video cassettes, records, kits, pictures, maps, etc.) the vast majority of the library materials budget goes to printed matter. In addition, the majority of the materials budget is spent on monographs with as little as 20% of the materials budget going to standing orders and subscriptions. This is in contrast to research libraries where up to 80% of the materials budget can be spent on standing orders and subscriptions. Many public libraries have acquired CD-ROM informational services and provide online data base searching services, but a very small percentage of the materials budget is used for these purposes. The vast majority of the materials budgets in public libraries is spent on monographs.

Nearly all medium-sized and large public libraries use a bibliographic utility (or a CD-ROM MARC data base) as their source of cataloging copy. They catalog according to AACRII conventions (as interpreted by the Library of Congress) and they prefer LC MARC records to non-LC MARC records. They classify nonfiction monographs using the Dewey Decimal Classification System and increasingly accept LC-suggested Dewey classification numbers. Fiction and periodicals are often not classified although they may (or may not) be cataloged. As a percentage of the total, relatively little original cataloging for monographic materials occurs in public libraries because cataloging copy is normally found in the bibliographic utility. However, non-print materials often require original cataloging because cataloging copy is not available.

Bibliographic access is in a state of change. Some public libraries continue to have multiple card catalogs and continue to produce and file massive numbers of catalog cards (a branch library system that catalogs 10,000 titles a year with an average of 5 copies/branches per title can easily generate 250,000 new catalog cards a year). Some have computer-produced union microform catalogs that may or may not include individual branch holdings. A few public libraries have CD-ROM catalogs and online catalogs but there are relatively few of these modern catalog forms actually operational in public libraries in 1990.

The percentage of staff and budget assigned to technical services in public libraries is generally smaller than in academic and research libraries while public services receives a significantly larger share of staff and budget resources. This is due to the nature of materials acquired and cataloged in a non-research public library (fewer titles, more copies, current trade English publications, more monographs/fewer serials, etc.) as well as the emphasis on public services in a public library (after all, public is in its very name!).

All public libraries are caught in the speed versus accuracy, quantity versus quality dilemma. Timeliness of availability of new titles is very important to these libraries. Unlike research libraries that are selecting, acquiring and cataloging for the ages, public libraries are selecting, acquiring and cataloging for the moment. Speed is all important. Public library users and the public services staff that serve these users want current fiction and non-fiction titles available within days of publication. Ideally, public libraries want new titles to appear on their shelves the same day they appear on book store shelves. Technical Services is thus caught in a dilemma—given limited staffing resources, is it better to acquire, catalog and process new titles quickly even if it means sacrificing purchase price discount and/or quality, accuracy and consistency of cataloging or is it better to emphasize discount, quality, accuracy and consistency at the expense of speed? Most public libraries make compromises both ways. Jobbers are sometimes selected based on speed of delivery instead of least price. Bibliographic short cuts are sometimes taken to get materials out on the shelves and accuracy and consistency are sometimes compromised. This reality is often rationalized by the fact that in many medium-sized public libraries the average life of a cataloged title is only three to five years because all copies are withdrawn (because the item is physically worn out or the information is no longer accurate or needed) or stolen. The bibliographic record is removed thus removing the inconsistency or inaccuracy from the library catalog.

Technical services employees in medium-sized public libraries are often isolated from the public services staff. In some cases technical services is located in a service center that does not have a public service component. The staff is often viewed as a production unit (i.e., factory) and is seldom invited to serve on library wide committees, etc. Often the public services staff believes the technical services staff is isolated and does not understand their problems and the technical services staff often feel the same way about the public services staff. This division and isolation is not universally true but it is more prevalent than we might want to think.

Automation in public libraries is in a great state of flux. Most small public libraries have little to no automation. However, the increasing availability of proven micro-based library software packages may gradually change this reality over the next several years. Almost all medium and large public libraries have automation activities within the library. For most of these libraries automation primarily means an online circulation/inventory control system to handle the millions of circulations that occur annually. Many medium-sized public libraries have had online circulation control systems for many years (ten and more) and they are an integral part of the day to day operations of the libraries and the technical services divisions that maintain them. Some public libraries have had online circulation control systems so long that they have had time to tire of their first system and have gone through the agony of changing systems to what is now referred to as "second generation systems." Still, for the vast majority of public libraries, the emphasis has been on circulation/inventory control, as opposed to bibliographic access, because this was the area perceived to be in crisis and thus in need of attention.

THE FUTURE

What about the future? What will be different? How will technical services in the public library change in the 1990's? What will technical services in the public library look like in the year 2000?

The next ten years are going to be enormously exciting for technical services in public libraries. While much will remain the same, a great deal of positive change is also going to occur. This change is going to be made possible by advances in technology and changing attitudes that are already in place.

Technology has advanced far beyond the minimum requirements of public libraries. Cost effective proven processing power, storage media, telecommunication networks and terminals exist in abundance. Application software to effectively utilize the technological power is being developed and will come to fruition in the 1990's. Public libraries will increasingly move from circulation/inventory centered automated systems to truly online bibliographic control and access systems with the emphasis changing from circulation/ inventory control to bibliographic control and access. The automated public access catalog will become preeminent rather than the automated circulation control system. Emphasis will be placed on information access with the marriage of the circulation and bibliographic systems so that public library users will have immediate and up-to-date access to information about what is available in the public library collections and the current status of each item in the collection. Truly comprehensive, integrated, user-friendly automated systems will increasingly be available from vendors encompassing bibliographic access, acquisitions, serials control, circulation control, and automated authority control. The effective integration of these "total systems" in the public library environment will be an exciting task for library managers.

In addition to advances in technological development, changes in attitudes are also occurring. The public library, while continuing to be viewed as a place to obtain educational and recreational materials (books, cassettes, etc.), will be increasingly viewed as a starting place to access current information on a variety of topics. Unlike educational and recreational materials which are largely contained within the collections of the public library, much of the information provided will come from external sources. While this distinction will hopefully be largely transparent to the user, it will have subtle but important impact on all parts of the public library including technical services. The public library will spend more of its resources on providing this "window" to information and correspondingly less of its resources on educational and recreational materials. These expenditures will involve personnel and the expenses of providing access to external information (long distance telephone calls, online external data base searching of both bibliographic and textual data bases, etc.). Much of the access to this external information is going to be provided via the library catalog and thus technical services is going to be intimately affected and involved. The subject catalog will be the most affected. It will include entries to external organizations and resources within the community that provide specialized information beyond the scope of the general public library. Because of its ease of accessibility (key word searching, etc.) and ease and speed of updating, (automated authority control, etc.) the catalog will become much more central and important to the services the public library provides.

By the year 2000, the advances in technology and changes in attitude already in place will have resulted in a changed technical services division at the public library. Yes, there will be a technical services division. Technical services will not disappear and somehow be absorbed within public services due to the specialized nature of the work, the required adherence to standards, and the production requirements. Maintenance of the catalog and physical requirements for acquiring and processing library materials will continue to make a central place for these tasks to occur the most cost effective alternative. But, the technical services division will be different. It may look largely the same — although card files will have been replaced by terminals. Technical services will continue to acquire catalog and process materials as quickly as possible to make them available for public use. But wise and judicious application of technology will allow the technical services staff in the year 2000 to be so efficient that less emphasis will be placed on production and more emphasis will be placed on providing access to the available materials in the public library collections and to available information and resources outside the walls of the public library. In short, the emphasis will move toward providing a truly open, user friendly, user responsive catalog.

Technical services and public services staff will work much more closely together in the task of maintaining the catalog. In 1990, maintenance of the catalog is viewed as a technical services task. In 2000, this will be viewed much more as a shared responsibility with· public services staff determining and communicating the changing needs of the public to technical services staff who will, with the aid of technology, be able to immediately respond to these changing needs. Additional user-friendly access points or references can be added within minutes of the need. By working together, the technical and public services staff will provide an infinitely more useful catalog (useful both to the public and the staff) than is provided in 1990.

The emphasis on having current materials in the public library will continue but additional emphasis will be placed on access to information much of which will be available externally. CD-ROM data bases (both indexes and textual materials) will be prevalent with access provided via the catalog. The catalog will contain access to the individual periodical articles that are contained within the library either by providing index entries in the local catalog data base that were purchased from commercial vendors or by providing access to an external index from the online catalog terminals. In addition, the catalog will contain references to specialized informational resources available in the community that are of general interest (e.g., art museum library, historical society library, etc.). For those users who desire to use it, the catalog will provide a link to the online catalogs of other libraries in the region, state, etc. so that both monographic and serial holdings of other libraries can be known.

These technological and attitudinal changes will have a direct impact on the professional staff in technical services. Their roles will be expanded and enhanced in a variety of ways. While they will continue to be concerned with meeting "production" goals so that new materials are available for public use quickly, they will also be directly involved in catalog data base maintenance activities that are intended not just to correct past errors and inconsistencies but more importantly are intended to increase the useability of the catalog for the public by adding additional access points and cross references far beyond those that are economically possible in manual catalogs. Staff will have to be more aware of informational resources available in the community so that references to these resources can be included in the online catalog data base. Technical services professionals will need to work more closely with their public service colleagues to insure the increasing useability and comprehensiveness of the catalog.

In order to accomplish these changes, technical services management will have to provide opportunities for their staff, and themselves, to increase their knowledge and skills in automation, communication and subject indexing. Technical services managers will need to be aware of local library/information networks in their geographic area so that they can provide linkages and interfaces to these networks. In addition, they will need to be knowledgeable about services that are available from regional/national networks and vendors that might be applicable and useful to their local automated library systems (i.e., the availability of indexes to periodicals that can be imported into their local data base). Most importantly, technical services managers will need to view themselves more as proactive change agents and less as reactive production managers.

In summary, technical services and public services will work together to provide a more useful and comprehensive catalog for the public to access. Technology will provide the tools to make this a reality (online bibliographic control systems with rich automated authority control systems including the capability for lots of notes in the catalog, etc.). Will this automatically happen? Of course not. Technological development can allow change to occur but it can't cause it to occur. People make change occur within an organization. Technological advances will provide tools for bright, energetic, risk-taking managers to make change in the way public libraries provide services. Management of technical services will be an important component in this process of changing service delivery. The technology is here. All that is needed to make change occur is committed, visionary, risk-taking managers. What an exciting time to be involved in technical services management.

The Role of Librarians
in Bibliographic Access Services
in the 1990's

Jennifer A. Younger

Bibliographic access services are derived from the functions of descriptive cataloging, subject cataloging, and classification. The principles and objectives used in the field today are, for the most part, derived from those established in the late nineteenth century, a brilliant period for librarianship in general and for bibliographic access in particular. The achievements of Melvil Dewey and Charles Cutter stand as living monuments to the profession's contribution to the classification and cataloging of documents.

A century later, we are in what will surely be another watershed period in the development of bibliographic access tools. Environmental and technological advancements of the 1970's and 1980's have laid the foundation through the implementation of machine-readable cataloging, shared cataloging, online catalogs and most recently, integrated access to library catalogs and journal indexes. Online searching functionality, search engines used to access multiple databases, linked systems, full text databases, electronic publishing, increases in computing power, and the promise of expert systems are in the process of forming a new technical platform on which to base revised methods of bibliographic access and control.

The purpose of this article is to identify the roles that professionals working in libraries should play in providing bibliographic access services and to suggest directions that libraries and schools of library and information studies can take to foster and support librar-

---

Jennifer A. Younger is Assistant Director, Technical Services at The Ohio State University, Columbus, OH.

125

ians in these efforts. *Professionals* means the staff with a masters degree in librarianship. When discussing the responsibilities of professionals in providing bibliographic access services, I refer to "professionals," but more often to "librarians" and "catalogers" (again on the assumption they hold a masters degree in librarianship) because, especially in the latter case, these titles are evocative of the more specific responsibilities. Some bibliographic access activities are described as examples of how these roles can and are being performed.

Approaching the next decade of bibliographic access in terms of roles is a conscious choice for reasons having to do with the importance of people and the nature of the conceptual framework under which they perform their responsibilities. It is axiomatic that individuals are the most important resource in any organization. They possess the knowledge, generate the ideas and carry out the work. How individuals define their responsibilities and activities is therefore a critical element in determining future advancements and progress of, in this case, bibliographic access services.

ROLE VERSUS TASK

Although the more often-discussed aspects of organizational structure are the design of positions and decision making systems, the focus here is on the conceptual or subjective framework that guides whether individuals view their responsibilities in the context of tasks or roles. The difference between these two approaches is not one of magnitude but rather one of the nature of the relationship of the individual to the organization. Tasks are static in nature and over time, the emphasis is on clarifying the precise content of that task. When the task content changes, the relationship to the organization is broken and must be initiated again through a redefinition of the tasks. In contrast, roles offer flexibility, ambiguity, and process, with the content being in more or less continual flux (Hurst 1984, 79).

Previous studies have compiled lists of bibliographic access tasks and described them as professional, technical or clerical in nature (American Library Association 1948; Ricking and Booth 1974). The disadvantages of this approach for describing professional responsibilities become apparent when, as is inevitable, the content and assignment of tasks to professional staff are changed. A case study reporting on the deprofessionalization of cataloging in six West Coast academic libraries revealed that the pessimism about the role and career expectations of catalogers was largely a result of changing task assignments (Hafter 1986, 64). Not only were computers dictating work assignments and schedules but more importantly, the acceptance of standardized cataloging copy left little original cataloging for professionals to do.

In this article, the term "roles" means the expected behavior of individuals as they go about performing their bibliographic access responsibilities. A conceptual framework that views professional responsibilities as roles both allows and encourages librarians to focus on the goal of providing effective bibliographic access services rather than the means used to get there. To the extent that activities are described, they are intended to be illustrative, not prescriptive, of professional responsibilities in providing bibliographic access services.

Roles for librarians both influence and are influenced by events taking place in the surrounding environment. Two influences are of particular importance to bibliographic access services, and because they are an integral foundation for the roles mentioned in this article, they are mentioned briefly: a sociological understanding of professional roles in librarianship and the impact of technology on information storage, transmission (including electronic publishing), and retrieval.

The role of librarian has evolved from "library keeper" to "library operator" to "librarian as a provider of professional information services" (Edwards 1975, 29). While the early emphasis was on controlling the use of the books and supervising production activities, the primary focus is on providing information services to clients. Within libraries, the responsibilities and duties have been redistributed among various levels of support staff and librarians, with the time of librarians reallocated to programmatic responsibilities. As defined by Veaner, programmatic responsibilities are those "intellectual choices which result in certain resources allocation decisions, decisions which form and drive the library programs supportive of institutional goals, objectives and academic programs"

(Veaner 1982, 6). Librarianship is seen as a gestalt that brings together complex functions focused on a goal — connecting users with
the information they desire — yet the means of accomplishing the
goals are constantly changing and adapting to new conditions. The
library's mission of selection, organization, and dissemination in
support of the information needs of the academic community remains of great importance yet at the same time the means by which
the mission is achieved will change.

Electronic publishing, at this stage just as difficult to forecast as
the impact of computers on bibliographic access, nevertheless will
be a catalyst in accelerating the kinds of changes described by
Edwards. Aveney forecasts (although by no means on any certain
timetable) that future decades will see a transfer of some library
operations functions, e.g., the storing and inventory functions, to
electronic bibliographic and text utilities (Aveney 1984, 74). Con-
currently, the demands for professional services in locating and
evaluating materials for library users will increase, a shift towards a
greater emphasis on public services. Publishers may also take a
greater role in furnishing cataloging data as part of the "online
shopping guide" necessary to let potential users know of the titles
they hold for on-demand publication (Brownrigg, Lynch and Engle
1984, 65). In the future, a great library may be defined by the quality of its bibliographic access tools as well as by its collections.

NINE ROLES PRESCRIBED

I propose nine roles for professionals in bibliographic access services: (1) providing leadership for bibliographic control activities in
the library as well as in the profession, (2) creating the bibliographic access system for the library, (3) coordinating bibliographic
access policies, (4) training, (5) managing a bibliographic access
department or system, (6) innovating, (7) boundary-spanning,
(8) evaluating the performance of the bibliographic access department or system, and (9) interpreting and conducting research. In all
of these roles, the common denominator is a thorough and comprehensive knowledge of the principles and means of creating bibliographic access together with the ability to communicate clearly and
succinctly. Whereas the responsibilities of individual positions are

based on the needs of the library and considerable variation is expected in the emphasis placed on one or more of these roles, it is the contention of this author that there is, nevertheless, some small amount of every role in most positions.

PROVIDING LEADERSHIP

Leadership is a role quite distinct from other professional roles such as the management of operations. Managers use their available resources to bring about specific accomplishments, while leaders focus on asking questions about why a particular operation exists at all and, in that light, offer guidance and direction focused on best accomplishing the task (Bennis and Nanus 1985, 21). As a process, leaders influence followers and, in turn, are influenced by them, both obtaining mutual goals. Of critical importance is the concept that leadership provides a focus through which organizations meet their goals and through which "followers can satisfy their personal needs while pursuing organizational goals" (Adams 1985, 35). The "great man" theory captures our imagination so that we often see only the small number of individuals who are seemingly larger than life, but in reality, leaders exist in all walks of life and at all levels in an organization. Leadership is a role calling for expertise and persuasive reasoning. That leadership is a role within the province of professionals with bibliographic access responsibilities is recognized by the several awards given annually by the Association of Library Collections and Technical Services (ALCTS).

Academic libraries are staggering under the burden of providing effective bibliographic access systems. While cooperative cataloging networks have ameliorated the growth and extent of cataloging backlogs, the adoption of new cataloging codes, new terminology employed as subject headings, preservation microfilming and retrospective conversion of catalog records originally done under now-outdated rules all make the continuing maintenance of catalogs a job of ever-increasing proportions. Under a cataloging ethos relying on consistency, complexity and quality cataloging (Hafter 1986, 129), the chasm separating administrators from catalogers is predictable yet regrettable.

Effective bibliographic access systems cannot depend on phenomena and values embedded in the status quo but must foster the emergence of new phenomena and values. New phenomena — MARC formats, *Anglo-American Cataloging Rules*, second edition, and uncontrolled subject descriptors — have appeared and been incorporated into the tools of the librarian's trade. Values, on the other hand, reflect a deeper level in librarians' belief system, the concern with what is important throughout time and changing circumstances. Professional values rest on service, a commitment to intellectual freedom, and preservation of the records of civilization. Service to the academic community is part of the mission statement as well as those in the bibliographic control community believe cataloging and classification to be significant contributions — culminating in the library catalog — to the overall mission of providing service.

Values are brought to everyday situations through statements of goals and objectives. In the last half of the nineteenth century, the stated objectives of library catalogs in the U.S. were to enable users to find specific titles as well as to find what the library had on a given subject and in a given kind of literature (Cutter 1904). Since then, however, the necessity of strict adherence to the values of Cutter has been a matter of some debate.

A growing body of evidence suggests that the majority of user needs could be solved through consulting a simple finding list of materials the library owns. Still unanswered, a fundamental question relating to the research objective of the catalog, and to a lesser degree to the finding list function, is "how much authority control is really needed?" (Svenonius 1981, 101). Wilson suggests the kind of power one would most like to have is power to get the best textual means to resolve one's particular goal (Wilson 1983, 15). He describes this as not the ability to discover everything there is that fits the bare topical description on the subject but rather something *good* to help her or him understand and solve the problem at hand. Further, in the eyes of some, the usefulness of the catalog has been eroded on yet another criterion. "The timeliness of bibliographic access tools has been effective in inverse proportion to the timeliness required of the information. Electronic information places a premium on immediacy of access to information, but our techniques for creating bibliographic tools are best equipped for the archival" (Graham 1990, 242). Catalog design based on values perceived as finding tools, evaluation, and timeliness would perhaps employ very different tools than we presently use.

It is the leaders in the field who must create a vision of what constitutes effective bibliographic access and, subsequently define the paths necessary to achieve effective bibliographic access. Just as new means emerge to replace the old, existing values must be reexamined and redefined in order to retain their effectiveness. Economic forces may initially propel new means of achieving bibliographic access, but unless the means of cataloging are related to values, they will not succeed.

That this is so can be seen by examining changes in higher education. In the 1870s radical changes occurred as land grant colleges were established and curricula designed to include science and technology, going beyond the classical education for the professions of teaching, medicine, law and the ministry (Kerr 1982, 163). These reforms succeeded because the entire country was industrializing and moving west of the Mississippi River. Higher education was reflecting value changes desired by university presidents, faculty, and the entire population. By contrast, some of the reforms attempted in the 1960s, including improved environments for undergraduates and integrated programs, failed because they moved in the direction of values that faculty members by and large opposed (Kerr 1982, 166).

A similar conflict in values is found with regard to minimal level cataloging, introduced as an economic necessity by administrators frustrated with cataloging backlogs and no prospects for their elimination. Assisted by technological factors of the sort that permit rationalization of these practices, including online catalogs that allow keyword access to title and contents data, minimal level cataloging has been forced into reality. Yet, in terms of acceptance by the community of bibliographic experts, it is doomed to failure because it does not rest on their underlying values of what bibliographic access should provide to users. Online catalog studies have indicated that the volume of subject searching conducted in library catalogs by far exceeds what was generally believed to be the case and, further, that such searching cuts across all categories of users, from beginners to disciplinary experts (Matthews, Lawrence, and Ferguson 1983). Catalogers have interpreted minimal level cataloging as inimical to user needs (which it is, although the tradeoffs are not inconsequential as far as creating a record of titles owned by a library), yet have themselves not responded with alternative methods for simplifying the process of cataloging, thereby reducing the cost. What is needed is strong leadership from within the community of bibliographic access experts that takes into account user needs, technology, and fiscal realities and defines a workable solution to the problem.

If bibliographic access services are to succeed in the next century, the introduction of new phenomena, technological and otherwise, must be securely based on underlying community values. Current methods of creating bibliographic access are rooted in the last century with its ideal of comprehensive searches. New values can be discovered, articulated, and incorporated but only if bibliographic access experts become leaders. Without arguing for the abandonment of present values, it is imperative that this generation of leaders focus on affirming the changed values needed in the context of contemporary information seeking behavior, disciplinary tools, searching capabilities, automatic indexing, expert systems, and other technological advances. Leaders in the field must also work in harmony with the fiscal realities of providing bibliographic access.

Academic libraries must expect such leadership and consider bibliographic access librarians a primary source of expertise in setting bibliographic access policies. When values dictate and support trends not in agreement with administrative efforts, bibliographic access librarians must lead the way in developing alternative solutions. In this context, a rejection of minimal level cataloging is insufficient. Bibliographic access librarians must lead us into the future by creating alternatives, for example, by redefining minimal level cataloging as cataloging requiring less time to complete but not as cataloging with fewer access points. Simplifying cataloging procedures, however, will happen only when bibliographic access librarians play a leadership role.

CREATING THE BIBLIOGRAPHIC ACCESS SYSTEM

Creating means bringing into existence. Cataloging rules and classification schemes are established, books and other materials are cataloged according to those rules, and the resulting records are brought together to form a catalog. In designing and creating the bibliographic access system, librarians perform a variety of specific activities, including establishing cataloging or indexing policies, creating specific cataloging and indexing tools such as thesauri, deciding the extent to which library materials will be cataloged and the priorities for doing so, and finally, performing original cataloging.

Since the days when budding librarians were taught "the library hand" as part of their library instruction, the tasks performed by catalogers have varied in accordance with the technology of the times. Shared bibliographic copy, available first with the *National Union Catalog* and greatly expanded by the bibliographic utilities, led to the introduction of paraprofessionals to handle the process of finding existing copy, matching it with the appropriate library materials and editing the copy in conformity with local policies and procedures. Paraprofessionals in many libraries also routinely create original descriptive cataloging records, forwarding them to a professional only for subject analysis. Although the existence of cataloging networks and paraprofessionals has left some professionals wondering what they are to do as catalogers, this stance only emphasizes the limitations of the task approach.

Despite the concern of some catalogers, there are professional activities in creating a bibliographic access system, although there is at the same time a recognition that many activities can be handled successfully by paraprofessionals. The distinction between the two sets of responsibilities is found in the nature of the question in hand. When a question is routine and readily interpretable in light of extant policies or of the piece in hand, paraprofessionals are well equipped to handle it. On the other hand, for questions of extreme complexity, as in titles requiring significant levels of investigation to establish bibliographic relationships or create access points, or questions requiring either the establishment or revision of policies, professional expertise is necessary to handle the question.

This calls for a formal team approach to cataloging in which professionals and paraprofessionals work in a close relationship. Teams formed on the basis of discipline or by libraries receiving the materials already exist with examples often seen in music, medical and law libraries. The advantages of this approach are several, and traditionally pertain to the ability to draw on the appropriate subject expertise in cataloging as well as to maintain contact with library users. From a cataloging perspective, there is another advantage and that is the opportunity to use staff expertise—both professional and paraprofessional—to the fullest extent. Rather than determining staff assignments on arbitrary criteria, such as whether the cataloging is original or copy, work flows and assignments could be designed to match the expected level of difficulty, routine and use of existing policies or complex and requiring new policies, with the appropriate level of staff expertise. The separation of cataloging activities into "functional" units, generally copy, original and authority control units, has in some part, fostered the sense of isolation felt by professionals as their original cataloging not only relates to lower priority library materials but also because of the quantity of copy cataloging, they may be only loosely connected to policy-making in areas such as bibliographic organization and authority control.

The role for librarians in creating bibliographic access was never defined solely by the volume of original cataloging to do in the library. As new formats of materials were introduced into library collections, e.g., audio and video cassettes, catalogers working at the local and national levels formulated the necessary modifications in cataloging rules and determined the extent to which these items should be cataloged. More recently, the computerization of bibliographic access activities has reintroduced an emphasis on catalog design. Online catalog construction has included activities focussed on index manipulation, data displays and command sequences, issues of catalog design that were settled many years ago for card catalogs.

On large university campuses, union card catalogs have been fairly common with varying degrees of coordination practiced in the cataloging policies. Nevertheless, machine-readable bibliographic records purchased from external sources such as the Government Printing Office (GPO) or the Center for Research Libraries (CRL), raise new questions of how to manage bibliographic data not created by the local library. Yet, in these cases, merging this externally-created bibliographic data into OPACs has merely extended, not altered, the basic premises of the catalog because the bibliographic data is created according to the same cataloging rules as is local data and the records represent materials owned by libraries.

Change of a more fundamental nature is on the horizon. Catalogers have had to become conversant with computers but the focus of their work has remained largely the creation of a local catalog. As the "library without walls" becomes a reality, creating bibliographic access to a particular library's collections will increasingly be done from the perspective of the user, which is to say that greater concern will be given to the use and interpretation of the catalog in light of other available tools.

In a card format, the library catalog was a discrete tool which, although housed in the reference department in close physical proximity to other indexes, was readily distinguishable from other printed indexes and interpreted largely in the context of cataloging rules. Now that online catalogs (OPACs) and other externally-produced online bibliographic databases can be accessed through the same work station, the domain of library cataloging will expand to include greater consideration for the relationship between library cataloging and journal indexing policies, especially in the areas of uniform headings, controlled vocabulary and free text, relationships among multiple thesauri, and the depth of indexing.

Adopting a broader perspective does not mean abandoning cataloging rules in favor of indexing rules, but rather adopting a greater level of awareness of how such rules work together to create a total system of bibliographic access. An early study of this nature examined differences in form between name headings established according to cataloging and bibliographic rules in an effort to understand how catalogs relate to other bibliographic tools (Tate 1963) while a later study concentrated on the impact of bibliographic and citing practices of conferences (East 1985). Tools such as the *Cross-Reference Index*, designed to link already-existing terms in discrete thesauri for the purpose of using varying terminology across multiple databases, while designed primarily for searchers, will ultimately be of use to catalogers who are developing new terminology for subject headings.

What is important for librarians to recognize is that the role of professionals in creating bibliographic access was always distinguished from that of other staff members due to their responsibilities for the establishment of cataloging tools, catalog design and policies in addition to cataloging individual items. Using the preferred model of librarianship described by Veaner, programmatic responsibilities, such as an analysis of the utility of uniform headings and the degree to which authority control should be implemented, can and must be continue to be distinguished from activities requiring the application of technical expertise, such as integrating new headings into an existing catalog (Veaner 1982, 13). Not only is the staff morale of paraprofessionals affected by fuzzy distinctions between the roles played by librarians and other staff (Kreitz and Ogden 1990) but also the design and creation of a bibliographic access system is stunted when deprived of professional talents. It is this area – the creation of a bibliographic access system – where a cooperative approach must be fashioned so as to maximize the distinct contributions of professionals and paraprofessionals.

COORDINATING POLICIES

Creating an effective bibliographic access system relies on standards and policies, the use of which is coordinated across the system. Consistency has been described as the "hobgoblin of small minds"; yet, in terms of user interpretation of the system, there is an important value inherent in establishing some degree of consistency.

Fifty years ago the coordination of cataloging was largely the job of a single department head with that department then performing the cataloging for the entire library. The increase in the numbers and formats of library acquisitions, unique disciplinary requirements, technology in the form of online processing systems, and an ecumenical approach to librarianship have resulted in the distribution of cataloging activities to many departments throughout the library: serials, government documents, law, music, cataloging, and reference. Among the benefits of decentralized cataloging is the ability to develop specialized expertise relating directly to the subject or type of material. The disadvantage is that departmental boundaries are barriers to communication, necessitating additional attention be given to coordinating policies across departments to which cataloging is distributed. This can be done successfully through a staff position, e.g., a bibliographic access coordinator, (Gleason and Miller 1987) in the same manner used to coordinate collection development policies.

PROVIDING TRAINING

Several texts directed toward on-the-job training in libraries outline the elements of an effective training program (Conroy 1978; Creth 1986). Although a library may have a training or staff development coordinator who coordinates the library's overall training program, there is still a role for bibliographic access librarians. Recognizing that the training role often overlaps with the supervisory role, it is discussed here separately because given the more complex organizational structures in libraries and the distribution of functions beyond central functionally-oriented departments, the provision of training extends beyond what supervisors do in single departments.

Training is discussed separately to emphasize as well its connection with leadership. More than anything else, training is an extension of leadership because it is leaders reaching out to enable staff—through training and information—to participate fully in library activities.

Bibliographic access librarians then, as supervisors or not, play a role in defining the content of training programs to meet the needs of individual library staff, including those who perform cataloging or who use the catalog in performing their responsibilities. This means identifying the bibliographic skills required for performing various responsibilities and, from their extensive knowledge of bibliographic access principles and local policies, planning the content of the training program. Setting up training also includes determining whether internal or external training, such as ALCTS regional institutes, is the most suitable for the situation.

Training is a labor intensive activity. Catalogers have another kind of opportunity to participate in local or national training programs. The development of computer-aided instruction for bibliographic control activities offers the dual promise of alleviating the personnel demands and delivering consistent training. A prototype computer-assisted instructional program is currently under development to offer training in the descriptive cataloging of computer software (Thomas and Weston 1990).

MANAGING

Managers direct the activities of other staff members and undertake the responsibility for achieving specified objectives through these activities (Katz 1974, 460). Putting aside issues such as budgeting for staff, supplies and in many cases, the purchase of bibliographic records from vendors, the focus here is on managing the activities of people. Even with this severe limitation, the surface can only be scratched in discussing the art of management.

The quality of management makes a difference in the performance of any department. A basic purpose of management is to support efficient production of what the department has the responsibility of doing. Skills in working with people — motivating, leading, communicating, negotiating — are of critical importance to managers. In the absence of such skills, respect and good relationships are difficult to develop and maintain. Department morale and productivity depend on good management.

Texts specific to managing catalog departments are useful to initiate neophyte managers on how catalog departments operate (Foster 1987), but more to the purpose are general texts on how to manage (Schein 1987) as well as texts on specific topics such as group processes (Ephross and Vassil 1988) or job enrichment (Hackman et al. 1975). In today's society, many individuals desire not mere employment but opportunities for contribution and satisfaction such that successful managers are moving away from the autocratic methods of past decades. The art of management is directed toward setting up relationships with staff that enable them to participate fully in determining and reaching the goals.

Empowerment, meaning literally to give power to someone, is a key concept in providing staff choice and control (Block 1987). Empowering staff to exercise choice and control begins with setting the overall direction and establishing the parameters within which choices can be made. It continues with providing the training, support and authorization for staff to proceed in deciding how to get the work done. Individual staff have some responsibility to be their own authority and, further, to state their commitment to the organization. Staff members must decide what they will do and how this will assist in achieving the organization's goals as well as their personal goals. Both are important to individual and departmental productivity.

INNOVATING

Innovation refers to the process for bringing new, problem-solving ideas into use. While one often thinks of innovations as primarily technological in nature, e.g., the use of microcomputers as cataloging work stations or the use of computer-assisted instruction, innovations are not limited to technology. Ideas for reorganizing the organizational structure or work flows, developing new budgeting systems and designing communication methods systems are also innovations. Creating better tools and methods is a major role in any organization and one that is readily accepted in the bibliographic access community. Since the adoption of call numbers indicating relative shelf location, the list of innovations has expanded steadily. Numerous examples exist, including many familiar practices once hailed as brilliant innovations: machine readable cataloging records, the *Anglo-American Cataloguing Rules (AACR)*, and reorganizations of technical services to take full advantage of automation.

Innovation begins with problem identification. A walk through any academic library or a brief literature search is a first step in locating problems, with further investigation necessary to determine its significance. The problem may relate to the operations of one library or be national in scope.

Organizationally-maintained roadblocks get in the way of innovation. The first step libraries must take is to recognize the organizational conditions under which innovations flourish, the second step clearly being to create those conditions. Three specific conditions promote innovation: information flowing freely across departmental (or other) boundaries, teamwork instead of dependence solely on hierarchical relationships, and resources devoted to new endeavors. Libraries can explore their "innovation potential" by identifying significant problems, putting a team into place, and providing them with the power tools — information, resources, and support (Kanter 1984, 357). The complexities inherent in innovation are illustrated through a brief recounting of one innovation that failed—a proposal by the Library of Congress (LC) to streamline shelflisting policies and procedures. In an effort to control the upward-spiraling costs of cataloging, LC proposed a change in shelflisting practices. If adopted, books would be arranged on the shelves first by the classification number and then chronologically by the date of publication rather than alphabetically by main entry. The response from the cataloging community was a deafening "No!" and the proposal died in infancy.

From a professional perspective, librarians cannot afford to handle innovative activities in this manner. There were indeed reasons why the proposed changes were not accepted (e.g., continuing need for arranging titles by author especially for literary and other creative works), yet the continued success of libraries rests in part on cost-saving innovations. A more appropriate response would have been the establishment of an ad hoc task force, including people knowledgeable about classification, shelflisting, the use of titles and editions by discipline, and browsing practices in libraries, with fiscal and administrative support for the necessary investigation. It seems entirely possible that the original proposal could have been examined more thoroughly in the context of varying user needs and subsequently amended so as to realize some streamlining (e.g., simplification of main entry arrangements) while preserving the status quo where necessary. Collectively, bibliographic access librarians missed an opportunity to find innovative methods of addressing the costs of shelflisting.

There is a last critical step to be considered retrospectively after the initial implementation has taken place. Library managements must work with innovators to determine the essential elements of the innovation that must be preserved and to report on the success and problems in the literature. In other words, it is important to recognize why something works, implement it accordingly, and inform the profession of the results.

BOUNDARY SPANNING

Uncertainty is a fundamental problem for complex organizations and coping with that uncertainty is an essential organizational process (Thompson 1967). Boundary-spanning literally means crossing borders and, in the case of bibliographic access services, refers to crossing the boundaries that separate cataloging from those who use the end result, both library staff and users. Crossing the boundaries is necessary to obtain information from the environment that will assist in defining how cataloging can be improved. As seen in the work of Cutter and Dewey, cataloging was a boundary spanning activity in the most perfect sense of creating cataloging rules in strict conformity to user needs. The principles of Cutter, after all, focused directly on providing cataloging entries of greatest convenience to the user and were reflected in his instructions, e.g., that the usage of the public, not of the cataloger, was to be the deciding factor in subject terminology (Cutter 1904, 69).

Yet cataloging is in danger of entering the next century as a white elephant. Rapid and profound environmental changes, of which keyword searching is the most easily observed and available example, have been largely uninterpreted and unused in the cataloging world for modifying cataloging rules. The isolation of catalogers, accomplished by placing them in departments located behind the scenes, was an organizational arrangement that established a stable environment in which cataloging expertise and production could flourish undisturbed by either library staff or the public-at-large. That isolation is counter-productive.

For understanding and responding to the environment, it is imperative that libraries view boundary-spanning as a critical element in carrying out bibliographic access functions. When faced with rapid and profound changes, organizations with more direct knowledge of the environment are at a distinct advantage in adapting old practices to new requirements (Kanter 1984, 41). Technology in the form of online catalogs has forced boundary spanning. Designing local interfaces, help screens, and search capabilities has brought catalogers into close working relationship with reference librarians, resulting more often than not in a better understanding on the part of all as to how cataloging influences information-seeking behavior and vice versa. What are needed are further efforts that bring catalogers into regular contact with external demands.

As currently practiced, cataloging depends very little on input at the local level, adhering rather to national cataloging rules and standards for successful performance. The relationship of cataloging standards to user needs by many accounts varies from predictive and close (e.g., the collocation of the works of major writers) to vaguely anticipatory and distant (e.g., information on three unnamed countries in South America). Escalating costs as well as the results of online catalog studies are driving factors demanding that cataloging be responsive to its current environment. Keyword searching capabilities have changed the nature of searching, yet cataloging practices have not responded with corresponding modifications. Without alteration, there is the potential that some bibliographic access practices in libraries will become obsolete in the face of online retrieval capabilities.

Within the library, boundary-spanning assignments for catalogers are being made with that response in mind. Recent organizational changes have brought catalogers into closer contact with users, often through integration of cataloging and reference (Busch 1986), sometimes through the involvement of catalogers in bibliographic instruction (Lawson and Slattery 1990). In both cases, the boundary-spanning occurs as catalogers bring a new understanding of user needs back to their cataloging responsibilities. From that more informed knowledge base, catalogers can, and do, modify extant cataloging policies and procedures (Younger 1990, 135).

Recent emphasis on the boundary-spanning nature of cataloging is seen in two review articles on bibliographic control and subject access (Svenonius 1990; Vizine-Goetz and Markey 1988). In reporting on classification research, Svenonius includes a section on research in user behavior and categorization of ideas in addition to the expected coverage of research on establishing bibliographical control systems. Vizine-Goetz and Markey examine the literature on subject searching in online catalogs as part of their review of the literature on subject access.

EVALUATING BIBLIOGRAPHIC ACCESS SYSTEMS

Evaluation has long been a concern and expressed role for bibliographic access librarians. Studies range from investigations of how to reduce cataloging costs (Dougherty and Heinritz 1982; Getz and Phelps 1984) to assessments of the use and effectiveness of the bibliographic access provided through the catalog by the descriptive cataloging rules, controlled vocabulary and classification schemes.

Assessing the impact and effectiveness of bibliographic access systems is more difficult than analyzing the efficiency of a particular operation. With the emphasis directed toward the impact on users, evaluation of a program, e.g., of minimal level cataloging would explore how well it was meeting the needs of library users. In so doing, it is necessary to distinguish various types of users, what kinds of materials they are looking for, what kinds of materials are receiving minimal level cataloging, the expected search keys, and the success of users in finding the items. Further, does this kind of program affect all users or is there one group that would receive greater benefit with less or no benefit to other groups?

Evaluative studies on the performance of catalogs cover many distinct topics such as the performance of uncontrolled versus controlled vocabulary. Online catalog transaction logs furnish a wealth of raw data for use in understanding how people are searching the catalog. Librarians are also approaching issues of subject access from disciplinary views and, in one set of studies, looking at the need for controlled vocabulary from the perspective of humanities scholars (Wiberly 1988). An exemplary study recently examined how well current cataloging rules for conferences serve the finding function of the catalog (McGarry and Yee 1990). Their conclusion was that the conference entries prescribed by the cataloging rules differed from what users would like to use.

The significance of these kinds of study can be seen in two arenas — national and local — because cataloging rules are established nationally and implemented locally. The next steps belong both to the authors of the article and to the bibliographic access community at large. The studies' results should be explored as a guide to possible modification of cataloging rules as well as of bibliographic instruction.

CONDUCTING LIBRARY AND INFORMATION STUDIES RESEARCH

Professional behavior depends on the presence of generalized and systematic knowledge (Barber 1965, 18) and research is the primary means of creating and extending such knowledge. Over time, professional practices can be expected to change based on knowledge developed through such study. Research is thus a basic obligation in which all professionals have a role, albeit at different points along the continuum as well as at different levels of intensity. Some individuals may focus on conducting research, others on using the findings to modify existing practices. The time devoted to carrying out such functions varies as well,with some individuals occupied solely with research while others handle it as a secondary activity. Accepting that all professionals have a role in research, the following suggestions are designed to foster their involvement: (1) use research in local library decision making and (2) handle major operational decisions as field experiments or studies. While the knowledge gained has the potential to improve existing methods, a single example illustrates the gulf that exists between research and practice. Years after a catalog use study indicated a chronological arrangement of subject cards would facilitate finding the most recent information in the catalog, many card catalogs continued to file subject cards alphabetically (Cochrane 1980, 109).

"To become operative in the world of practice, research results need to become known and established to the point that large numbers of practitioners accept them as self evidently superior to those assumptions they now accept" (Hewitt 1990). Current research in bibliographic control functions is regularly reported in journals including the *Journal of the American Society for Information Science*, *Cataloging & Classification Quarterly*, *Library Resources & Technical Services*, the *OCLC Annual Review of Research*, as well as in dissertations and monographs. Publication, however, does not guarantee use.

Academic libraries' support for research often comes in the form of release time or funding for computing and printing costs. Yet, in the view of one library administrator, a more significant contribution toward the enhancement of library research would be made if library administrators "make decisions and develop policies which work toward the integration of research into the operation of library" (Hewitt 1990). One simple measure to get the process underway is to require that the analysis of local problems incorporate relevant research in developing solutions. With regard to authority work, for example, an often-asked question on the local scene is whether to put cross references to preferred forms of names and subjects in the online catalog. Several studies on the success of catalog users in finding the preferred terminology could, if consulted, inform local decision making on catalog design (Bates 1977; Carlyle 1989; Taylor 1984). In making the assimilation of new knowledge—some of which will certainly challenge existing assumptions—a formal requirement of decision making, libraries would be providing the stimulus to encourage research to influence library practice.

Beyond the use of extant research, libraries can take steps to promote research within the library by carrying out major operational changes as field experiments or studies. As with any type of scientific research, these studies aim to discover relationships among variables, but they differ from other methods in that they are conducted in a real life setting. Field experiments and studies are further distinguished from each other by the amount of control that can be exercised over the situational variables—field studies lack the possibility of control. The imposition of a systematic process for describing existing conditions, stating assumptions, predicting outcomes, and determining the criteria and measures for evaluation strengthens the results. A case in hand is the implementation of the *Anglo-American Cataloguing Rules*, second edition. Extensive planning preceded the implementation but fourteen research libraries reported the implementation was considered successful or not largely on *post hoc* analysis of whether it appeared to run smoothly, with few questions from reference staff, or whether cataloging departments were able to maintain a respectable cataloging output (Hopkins and Edens 1986). How much more powerful these reports would be with the addition of specific predicted outcomes, evaluation criteria, and measures for determining success. While observations were made (such as that the process of following through on a large number of cross references made searching less efficient (Hopkins and Edens 1986, 148), no data was collected that would support anything other than personal observations on how often references were encountered and used in searching. Conclusions about the impact on catalog users and the usefulness of the time spent making cross references remain tentative.

Conducting and using research in the operational environment requires research skills appropriate to performing as well as interpreting research. Schools of library and information studies have incorporated, or are now incorporating courses on research methods into their curricula. Without slighting the need for skill in conducting research, clearly the ability to interpret research should be a basic requirement. When research findings are available as one factor in decision making, together with intuition, fiscal constraints and political realities, the ability to interpret research and to apply it to the situation at hand is critical. A dialogue among the intellectually curious offers a cross fertilization completely appropriate to research aimed at finding answers to practical problems as well as testing theories (Svenonius 1981; Svenonius 1986). Research findings unheeded in the field may result from research questions having little utilitarian value, although it is important to keep in mind that not all research should carry practical implications. The Council of Library Resources sponsors partnerships between researchers and practitioners, who may be one and the same but more often are not, offering another example of how research skills may be joined with ideas generated from the field.

CHALLENGE AND PROFESSIONALISM IN JOBS

Professional responsibilities have been defined in the context of roles with examples given of the kinds of activities that can be performed in carrying out those responsibilities. Recent articles and symposia have lamented the shortage of qualified trained catalogers, but not much has been written about what kinds of jobs are out there waiting for them. A recent issue of *College & Research Libraries News* (May 1990) listed several openings for catalogers. The libraries were seeking librarians to catalog books and, in one case, manuscripts, in a wide variety of subjects and languages. A major concern is the lack of responsibility for performing any role other than creating original cataloging. Positions of assistant department head and department head did, however, list responsibilities for policy development. A position offered under the title Information Access Librarian for Bibliographic Control offered no stated opportunities for participating in decisions about the programmatic direction of bibliographic control, but did offer the opportunity to participate in collection development and serve as liaison to academic departments.

The concern here is not that cataloging should not be handled by professionals, but rather that the positions available, as described in the job announcements, do not specifically indicate broader roles or responsibilities of the kind just discussed. Strenuous arguments have been made that programmatic responsibilities differentiate professionals from other library staff (Veaner 1982, 6). If this argument is accepted, then the absence of such responsibilities is troubling. One has to ask whether such positions are dead end jobs leaving little room for growth or advancement once the technical skills have been mastered through experience. After the intricacies of cataloging are learned, what do these jobs offer to challenge professionals? One possible explanation for why cataloging positions are going begging is because they are not described or viewed as challenging or exciting positions, and from that view, the negative stereotypes of cataloging positions are unlikely to change. If, on the other hand, libraries are indeed creating challenging positions for catalogers, the position announcements should be modified to reflect the challenges and opportunities awaiting the professional.

CONCLUSIONS

Professional roles in providing bibliographic access services are complex and challenging. The information explosion is being followed by the information sources confusion, while the need for an effective bibliographical structure is greater than ever to connect users with the information and materials they are seeking. Bibliographic access librarians must assume a leading role in bringing the profession into a new era of bibliographic access and control. Designing, creating and coordinating a bibliographic access system that will continue into the next century requires a team approach to utilize the expertise of both professionals and paraprofessionals. Training, managing, innovating, boundary-spanning, evaluating and conducting research are roles considered integral to professional responsibilities in providing bibliographic access services.

The responsibilities in each of these roles were purposefully described with selected examples in the full knowledge that many other activities and examples exist.

Though sometimes stuck behind the doors of technical services divisions, bibliographic access professionals are directly engaged in providing service to library users and their mission—creating an effective bibliographic access system—is vital to the success of users. The achievements of the past century together with the promises of the future point to new challenges and opportunities in providing bibliographic access services.

SELECTED REFERENCES

Adams, Jerome and Janice D. Yoder. *Effective Leadership for Women and Men*. Norwood, NJ: Ablex, 1985.

American Library Association. Board on Personnel Administration. Subcommittee on Analysis of Library Duties. *Descriptive List of Professional and Non-Professional Duties in Libraries*. Chicago: American Library Association, 1948.

Aveney, Brian. "Electronic Publishing and Library Technical Services." *Library Resources & Technical Services* 28(1984):68-75.

Barber, Bernard. "Some Problems in the Sociology of the Professions." In *The Professions in America*, edited by K. S. Lynn and the editors of *Daedalus*, 15-34. Boston, Houghton Mifflin, 1965.

Bates, Marcia. "Factors Affecting Subject Catalog Search Success." *Journal of the American Society for Information Science* 28(1977):161-69.

Bennis, Warren and Burt Nanus. *Leaders: The Strategies for Taking Charge*. New York: Harper & Row, 1985.

Block, Peter. *The Empowered Manager: Positive Political Skills at Work*. San Francisco: Jossey-Bass, 1987.

Brownrigg, Edwin, Clifford Lynch and Mary Engle. "Technical Services in the Age of Electronic Publishing." *Library Resources & Technical Services* 28(1984):59-67.

Busch, B. J. *Integration of Public and Technical Services Functions*. Washington, DC: Association of Research Libraries, 1986.

Carlyle, Allyson. "Matching LCSH and User Vocabulary in the Library Catalog." *Cataloging & Classification Quarterly* 10(1989):37-63.

Cochrane, Pauline A. "Catalog Users' Access from the Researcher's Viewpoint: Past and Present Research Which Could Affect Library Catalog Design." In *Closing the Catalog*, edited by D. K. Gapen and B. Juergens, 105-22. Phoenix: Oryx Press, 1980.

Conroy, Barbara. *Library Staff Development and Continuing Education*. Littleton, CO: Libraries Unlimited, 1978.

Creth, Sheila D. *Effective On-the-Job Training: Developing Library Human Resources*. Chicago: American Library Association, 1986.

Cutter, Charles A. *Rules for a Dictionary Catalog*. 4th ed. Special Report on Public Libraries, pt.2. Washington, DC: Government Printing Office, 1904.

Dougherty, Richard M. and Fred J. Heinritz. *Scientific Management of Library Operations*. Metuchen, NJ: Scarecrow Press, 1982.

East, John W. "Citations to Conference Papers and the Implications for Cataloging." *Library Resources & Technical Services* 29(1985):184-94.

Edwards, Ralph M. *The Role of the Beginning Librarian in University Libraries*. Chicago: American Library Association, 1975.

Ephross, Paul H. and Thomas V. Vassil. *Groups that Work*. New York: Columbia University Press, 1988.

Foster, Donald L. *Managing the Catalog Department*. 3d ed. Metuchen, NJ: Scarecrow Press, 1987.

Getz, Malcolm and Doug Phelps, "Labor Costs in the Technical Operation of Three Research Libraries." *Journal of Academic Librarianship* 10:209-19.

Gleason, Maureen L. and Robert C. Miller. "Technical Services: Direction or Coordination?" *Technical Services Quarterly* 4(Spring 1987):13-9.

Graham, Peter S. "Electronic Information and Research Library Technical Services." *College & Research Libraries* 51(May 1990):241-50.

Hackman, J. Richard, Greg Oldham, Robert Janson, and Kenneth Purdy. "A New Strategy for Job Enrichment." *California Management Review* 17(1975):57-71.

Hafter, Ruth. *Academic Librarians and Cataloging Networks: Visibility, Quality Control and Professional Status*. Contributions in Librarianship and Information Science, no. 57. New York: Greenwood Press, 1986.

Hewitt, Joe A. "The Role of the Library Administrator in Improving LIS Research." Paper presented at the annual meeting of the American Library Association, Chicago, June 1990.

Hopkins, Judith and John A. Edens, eds. *Research Libraries and Their Implementation of AACR2*. Greenwich, CN: JAI Press, 1986.

Hurst, David K. "Of Boxes, Bubbles and Effective Management," *Harvard Business Review* (1974):78-88.

Kanter, Rosabeth Moss. *The Change Masters: Innovation & Entrepreneurship in the American Corporation*. New York: Touchstone, 1984.

Katz, Robert L. "Skills of an Effective Administrator." In *Management Strategies for Libraries*, edited by Beverly P. Lynch, 459-80. New York: Neal-Schuman, 1985.

Kerr, Clark. *The Uses of the University*. 3d ed. Cambridge, MA: Harvard University Press, 1982.

Kreitz, Patricia A. and Annegret Ogden. "Job Responsibilities and Job Satisfaction at the University of California Libraries." *College & Research Libraries* 51(1990):297-312.

Lawson, V. Lonnie and Charles E. Slattery. "Involvement in Bibliographic Instruction among Technical Services Librarians in Missouri Academic Libraries." *Library Resources & Technical Services* 34(1990):245-8.

Matthews, Joseph, Gary Lawrence and Douglas Ferguson, eds. *Using Online Catalogs*. New York: Neal Schuman, 1983.

McGarry, Dorothy and Martha M. Yee. "Cataloging Conference Proceedings." *Library Resources & Technical Services* 34(1990):45-53.

Ricking, Myrl and Robert E. Booth. *Personnel Utilization in Libraries: A Systems Approach*. Chicago: American Library Association in cooperation with Illinois State Library, 1974.

Schein, Edgar H., ed. *The Art of Managing Human Resources*. New York: Oxford University Press, 1987.

"Retrieval of Selected Serial Citations." *College & Research Libraries* 50(1989):532-542.

Svenonius, Elaine. "Bibliographical Control." In *Academic Libraries: Research Perspectives*, edited by M. J. Lynch and A. Young, 38-66. Chicago: American Library Association, 1990.

Svenonius, Elaine. "Directions for Research in Indexing, Classification, and Cataloging." *Library Resources & Technical Services* 25(1981):88-103.

Svenonius, Elaine. "Unanswered Questions in the Design of Controlled Vocabularies." *Journal of the American Society for Information Science* 37(1986):331-40.

Tate, Elizabeth L. "Main Entries and Citations." *Library Quarterly* 33(April 1963):172-91.

Taylor, Arlene. "Authority Files in Online Catalogs: An Investigation of Their Value," *Cataloging & Classification Quarterly* 4(1984):1-17.

Thomas, Sarah E. and Claudia V. Weston. "CatTutor for Catalogers at National Agricultural Library." Paper presented at the annual meeting of the American Library Association, Chicago, June 1990.

Thompson, James D. *Organizations in Action*. New York: McGraw-Hill, 1967.

Veaner, Allen B. 1982. "Continuity or Discontinuity—A Persistent Personnel Issue in Academic Librarianship." In *Advances in Library Administration and Organization*, edited by G. B. McCabe, B. Kreissman, & W.C. Jackson, v. 1, 1-20. Greenwich, CT:JAI Press.

Vizine-Goetz, Diane and Karen Markey. "Subject Access Literature, 1987." *Library Resources & Technical Services* 32(1988): 337-51.

Wiberly, Stephen E. "Names in Space and Time: The Indexing Vocabulary of the Humanities." *Library Quarterly* 58(January 1988):1-28.

Wilson, Patrick. "The Catalog as Access Mechanism: Background and Concepts." *Library Resources & Technical Services* 27(1983):4-17.

Younger, Jennifer A. "University Library Effectiveness: A Case Study of the Perceived Outcomes of Structural Change." Ph.D. diss., University of Wisconsin-Madison, 1990.